STURGEONS IN AUSTRIAN RIVERS: HISTORIC DISTRIBUTION, CURRENT STATUS AND POTENTIAL FOR THEIR RESTORATION

Thomas Friedrich

University of Natural Resources and Life Sciences, Vienna
Institute of Hydrobiology and Aquatic Ecosystem Management
Max - Emanuelstraße 17
1180 Vienna, Austria

World Sturgeon Conservation Society: Special Publication n° 5 (2013)

World Sturgeon Conservation Society: Special Publication Series
Series Editor: Harald Rosenthal (President WSCS)
Special Publication n° 5

Sponsors

Austrian Fishing Association

Upper Austrian Fishing Association

Consultants on fisheries and aquatic ecosystems

Frontpage:
Russian sturgeon (*Acipenser gueldenstaedtii*) (top);
Fragementation of important sturgeon river stretches in Austria (bottom, left));
Stocking of Russian sturgeon in the Austrian Danube (bottom, right)

Editing and pagination by P. Bronzi

Publisher and produced by BoD - Books on Demand, Norderstedt, Germany
ISBN 978-3-732231975

Contents

Foreword

Worldwide most sturgeon species are either endangered or at the brink of extinction. Only a very few populations are managed sustainably. There are many anthropogenic factors responsible for this dramatic decline, including river fragmentation (resulting in longitudinal and lateral disruption), and loss of habitat. Other impacts are overfishing, illegal fishing and pollution. Sturgeons are commonly called *"living fossils and endangered giants"*. Although very old, our knowledge on their history and fate in many regions is extremely fragmentary. In many areas we depend on anecdotal information on their distribution, abundance and population status. Specific efforts and unconventional approaches are needed to recover existing information that is burried in traditional knowledge of local people or is based on occasional and rare records.

In the Danube River basin five anadromous and potamodromous sturgeon species are native, which partially used to migrate upstream the river to finally reach the Bavarian Danube for spawning. The exact distribution of the various species within the Danube River basin still remains unclear. In the Austrian section only a small population of the potamodromous sterlet (*Acipenser ruthenus*) is presently left, but threatened with extinction.

In the last few years sturgeon stocking (both accidentally and deliberately) and catches have increased throughout Austria. Unfortunately, many of these sturgeons are allochthonous species that pose a threat to the autochthonous sterlet population. It is imperative that steps are being taken to protect and support the remaining populations of sterlet. At the same time other stretches should be assessed as to their potential of supporting a viable sterlet population in Austria. Measures for re-introduction should be well-prepared and closely monitored as previous stocking programmes did not have a significant impact on catches in most cases and almost never led to the establishment of self-reproducing populations.

The objective of this study was to search for all available data and informations on sturgeons in Austrian waters to obtain a clearer overall understanding of their historical and current distribution in this country. The intentional use of the gained data is to evaluate the potential for Austrian rivers to support viable sturgeon populations again.

We hope this study can provide an incentive for neighbouring countries to undertake similar surveys in order to get a better picture of the fate of sturgeons in the entire Danube watershed at least during the past few decades. The WSCS supports these initiatives and, through its special publication series, provides a platform for disseminating the survey results.

June 2013

Harald Rosenthal
World Sturgeon Conservation Society
Neu Wulmstorf, Germany

Thomas Friedrich
Institute of Hydrobiology and Aquatic Ecosystem Management
Vienna, Austria

Paolo Bronzi
World Sturgeon Conservation Society
Monza, Italy

Acknowledgements

I sincerely thank my parents, who did not only support me during the study period, but - over the years - also became sturgeon fanatics. I am also indebted to my supervisor   Professor Dr. Mathias Jungwirth for his continued support and full trust in me to serve the objectives of the study.

Also many thanks go to various other persons providing data and pictures while answering many questions and reviewing material These include: Hans-Peter Angerer, Tim Cassidy, Josef Dellinger, Jürgen Eberstaller, Robert Elsbacher, Martin Friedrich, Michael Friedrich, Georg Fürnweger, Clemens Gumpinger, Werner Gritsch, Gertrud Haidvogl, Birgit Haring, Martin Hochleithner, Wolfgang Honsig-Erlenburg, Lucia Jirku, Regine Jungwirth, Nicolai Kogler, Rudolf Kovarik, Harald Kromp, Heinz Machacek, Matthias Maier, Ernst Mikschi, Günther Parthl, Wolfgang Petrouschek, Alfred Pleyer, Siegfried Pilgerstorfer, Clemens Ratschan, Heinz Renner, Klaus-Jürgen Rudowsky, Walter Salzmann, Ursula Scheiblechner, Bernhard Schmall, Stefan Schmutz, Michael Schremser, Christoph Trost, Herwig Waidbacher, Helmut Wellendorf, Franz Wiesmayer, Christian Wiesner, Gerald Zauner, Bernhard Zens.

Special thanks go to Professor Dr. Harald Rosenthal, Dr. Paolo Bronzi and the World Sturgeon Conservation Society for editing and publishing this study as a special publication of the WSCS series.

1. Introduction

Sturgeons are an ancient order of fish (Acipenseriformes), dating back in their occurrence for over 200 Million years. The order comprises two families (Acipenseridae and Polyontodidae) and has in total 27 species. They are naturally restricted to the northern hemisphere. Sturgeons exhibit a very long life cycle (maximum lifespan up to over 150 years; depending on species). They are late maturating species and many grow to very large sizes (up to 6 - 7 meters). Most of the sturgeon species are anadromous. There are also potamodromous (landlocked) species and forms, spending their entire life cycle in freshwater. Worldwide, various sturgeon species are already extinct, missing, highly endangered or vulnerably threatened through human impacts such as overfishing, damming, habitat destruction and pollution. Obstacles along migratory routes are among the major threats and prevent recovery efforts in many river systems. In addition the global demand for caviar exacerbates this trend, being an incentive for illegal fishing.

In the Austrian section of the Danube River five sturgeon species are native: two are potamodromous (*Acipenser ruthenus* and *Acipenser nudiventris*) and two are anadromous species (*Acipenser stellatus* and *Huso huso*) while an additional species (*Acipenser gueldenstaedtii*) exhibits both an anadromous and probably a potamodromous form. Heavy overfishing of the three large anadromous and one large potamodromous species in earlier centuries already contributed to their extinction in Austria. The construction of various hydrodams in the 20th century finally prevented their spawning migration (especially the construction of the Djerdap I and II dams at the Iron Gate in 1972 and 1984) and altered the hydromorphology and associated habitats along the entire river system. In the medieval period fences were constructed across the whole river width in order to block the migration routes and easily catch the fish. This fishing method proved so effective that it took only approximately two centuries for the sturgeons to rapidly decrease and nearly vanish in Austrian waters. Today one small freshwater species, the sterlet (*Acipenser ruthenus*), can still be found in our waters, but only in small quantities and in restricted areas. The populations of the remaining species are still threatened and their existence often depends on stocking. To this extent there is neither sufficient knowledge about the historical distribution of the various sturgeon species within Austrian rivers, nor about the actual population status as well as the habitat use of sterlets. For the remaining populations, increased stocking and catches of native and non - native sturgeon species might add further pressure through hybridization and competition. Intensification of sturgeon aquaculture will further increase this trend.

The objective of this study was to collect and compile as much information on acipenserids from Austrian rivers as possible. Four steps have been undertaken to accomplish this task: (1) various historical data and reports have been screened and analyzed to obtain a better picture on the historical distribution of the species within the Danube river basin; (2) to examine the current situation and stock status, using catch data from the last 30 years while also compiling stocking records (through interviews of experts, fishermen and scientific organizations);

(3) to attempt a link of the data from (1) and (2) with hydromorphological information to iden-tify the location of areas suitable for sturgeon conservation and potential reintroduction; and (4) to assess the extensive pertinent literature in an attempt to identify existing obstacles and potential measures for successful protection and reintroduction.

2.   Material and methods

The first part of this study considers the historical distribution of sturgeons within Austrian wa-ters. Four steps were involved in the study design: (a) a literature survey, going through various historical texts, books and other documents; (b) screening of over one hundred sturgeon preparations from the archive of the Museum of Natural History of Vienna, analyzing their spe-cies identity and the location of their catch; (c) creation of a database (Historic Database - HDB) and maps for the different sturgeon species (Page 29 - 32) based on the information gained through step (a) and (b), while also distinguishing actual catch reports and general information retrieved from various sources; and finally (d) compare the created database with information

Figure 1: Sterlet preparate at the Museum of Natural History in Vienna, Austria (Foto: Author)

from other external sources in order to ensure a coverage of historic knowledge as compre-hensive as possible. (Fig. 1)

To determine the current status of sturgeons in the area it was inevitable to look for in-formation on recent catches and stocking activities to create a database (Present Database - PDB). Consequently more than 300 organizations, governmental institutions, private persons, fish ecologists, fishermen, scientific organizations, sport fishing clubs and others were con-tacted. "Wanted posters" were placed in all fishing shops in Vienna and Linz, printed in fishing magazines as well as digital copies on various internet platforms (Annex II). Despite this effort, the final list of catches on stocking actions cannot be claimed to be complete. However, based on this information, the species and origin of both stocked and caught fish was determined. Catches in various lakes, ponds, and other water bodies were excluded as many of these have been deliberately stocked with various sturgeon species and thus their inclusion would gener-ate biased results compared to the situation in rivers. If there was no picture available and the report was not from an institution qualified to distinguish between the various sturgeon species, the specimens would be recorded as "unknown identity" in the database. Most fish were reported as "sterlets" or "sturgeon" by fishermen, using only the length of the snout as indication. Therefore many caught "sterlets" were finally identified as Siberian sturgeons by this author. Specimens caught before 1994 were generally considered to be sterlets, as the anadromous species were already extinct by then and sturgeon trade and aquaculture didn´t commence with stocking and releases until the mid- to late nineteen-nineties.

Using the two databases (HDB & PDB), the potential of various river stretches in Austria were assessed regarding suitability for sturgeon restoration programs. Because there is little to no knowledge on habitat use of sturgeons in Austrian rivers a very basic approach had to be used. The criteria to identify potential habitats were: (a) historic and current occurrences, (b) condition of caught fish, (c) signs of spawning activity, (d) length of available river stretches/ impoundments, (e) length of connected tributaries, (f) fragmentation and (g) habitat heterogeneity. Functional fish passes for sturgeons are still in the early development stages and in the foreseeable future it is unlikely that such passes will be installed in Austrian rivers. Furthermore the problem of downstream migration is still unresolved. Therefore each impounded section had to be evaluated separately. As a result a map was created identifying hotspots for sturgeon restoration (Fig. 18). The outcomes of this evaluation were discussed after an extensive literature study.

3.   Sturgeon Species in the Danube River Basin

3.1.   Native species

Five sturgeon species are native in Austrian waters. One freshwater species, the sterlet (*Acipenser ruthenus*) can still be found in small quantities. In the Danube it starts spawning with 3 - 5 years (males) and 4 - 8 years (females). Reproduction takes place every 1 - 2 years, the spawning migration in the river ranging up to 300 km (HOLCIK, 1989). Spawning time is in spring and the spawning places have similar characteristics as for other Danube sturgeons, like high flow velocity, water depths of 2 - 15 meters and gravel or rock substrates. It is also often sold in hatcheries for ornamental purposes; especially the white albino form is very popular for garden pond owners. The potamodromous ship sturgeon (*Acipenser nudiventris*) is nowadays limited to the middle Danube and on the brink of extinction as few specimens are rarely caught. Also there is no programme for controlled propagation of the Danube stock as no brood stock is available. The Russian sturgeon (*Acipenser gueldenstaedtii*), the stellate sturgeon (*Acipenser stellatus*) and the beluga sturgeon (*Huso huso*) are now restricted to the lower part of the Danube, as the dams at the Iron Gate block the upstream migration. Within the Danube two different migration types are known to exist for at least the beluga sturgeon and the Russian sturgeon: the vernal form migrates and spawns in spring, and the hiemal form migrates during the fall, overwinters in the river, and finally migrates further upstream to spawn in the following spring (KHODOREVSKAYA et al., 2009). All three species are also highly endangered, in particular the long- distance migratory hiemal forms. For all three species controlled propagation and stocking programmes are carried out in the lower Danube. These species can also be found in hatcheries and aquaculture facilities in Austria, however in most cases it is not clear whether the fish belong to Danubian or Caspian genotypes. Consequently, introduction of these fish into natural water bodies has to be considered as a potential risk to the autochthonous species. There is an ongoing discussion about a resident form of the Russian sturgeon in the middle Danube (HECKEL & KNER, 1857; HOLCIK et al., 1981; HENSEL & HOLCIK, 1997). If such a form existed it would probably be already extinct, as catches within the last twenty years (GUTI, 2006) are more likely to be fish that have escaped from hatcheries, rather than indicators for a relict population. Another sturgeon species, the common European sturgeon (*Acipenser sturio*) occurred historically only in the lower Danube and did not reach the Austrian section (BACALBACA- DOBRIVICI & HOLCIK, 2000).

3.2.   Conservation status

Sterlet (*Acipenser ruthenus* LINNAEUS, 1758)
  Red List Austria (MIKSCHI & WOLFRAM, 2007): critically endangered, estimated number of reproducing adults < 1000
  Habitat Directive (EUROPEAN COUNCIL, 1992): Annex V
  CITES (CITES, 2012): Annex II
Ship sturgeon (*Acipenser nudiventris* LOVETSKY, 1828)
  Red List Austria (MIKSCHI & WOLFRAM, 2007): regionally extinct
  Habitat Directive (EUROPEAN COUNCIL, 1992): Annex V
  CITES (CITES, 2012): Annex II
Russian sturgeon (*Acipenser gueldenstaedtii* BRANDT, 1833)
  Red List Austria (MIKSCHI & WOLFRAM, 2007): regionally extinct
  Habitat Directive (EUROPEAN COUNCIL, 1992): Annex V
  CITES (CITES, 2012): Annex II
Stellate sturgeon (*Acipenser stellatus* PALLAS, 1771)
  Red List Austria (MIKSCHI & WOLFRAM, 2007): regionally extinct
  Habitat Directive (EUROPEAN COUNCIL, 1992): Annex V
  CITES (CITES, 2012): Annex II
Beluga sturgeon (*Huso huso* LINNAEUS, 1758)
  Red List Austria (MIKSCHI & WOLFRAM, 2007): regionally extinct
  Habitat Directive (EUROPEAN COUNCIL, 1992): Annex V
  CITES (CITES, 2012): Annex II

3.3.   Alien species

For economic reasons various exotic sturgeon species and hybrids have been introduced into Austria. The most important one is the Siberian sturgeon (*Acipenser baerii*) which is produced in many countries of the northern hemisphere for its caviar and meat. Due to its robustness and lively behaviour it is also often sold for ornamental or sport fishing purposes. It can be

Figure 2: Native sturgeon species of the Danube River basin ((a,b,d): Author; (c,e): Clemens Ratschan)

found in many hatcheries, pet stores, etc. and is often confused with the sterlet. The white sturgeon (*Acipenser transmontanus*), originating from the North American Pacific coast, is produced mostly in Italy for caviar production and the males are sold throughout Europe for ornamental and especially sport fishing purposes. This species grows fast but is not as easy to handle as the Siberian sturgeon. There have been some viral diseases with high mortalities in hatcheries and occasionally in the wild in the United States (VAN EENENNAAM et al., 2004) which, if transferred, can also infect European sturgeon species (HOCHLEITHNER, 2004). Also imported, but in smaller numbers, is the Adriatic sturgeon (*Acipenser naccarii*), mainly for ornamental and "collector" purposes. In the last years also small numbers of Atlantic sturgeon (*Acipenser oxyrinchus*) became available on the ornamental market. The North American paddlefish (*Polyodon spathula*) is mainly produced in Eastern Europe, and was long thought to be an interesting species for polyculture with carp. The production numbers, however, decreased in the last years. In addition to pure species various sturgeon hybrids have been imported, the most important being the bester (*A. ruthenus* x *H. huso*), the osster (*A. ruthenus* x *A. gueldenstaedtii*) and the "AL" (*A. naccarii* x *A. baerii*). All these species pose a potential threat to autochthonous (sturgeon) species once introduced into natural water bodies, either through hybridization with native species, predation or competition for food and habitat. (Fig. 3)

(a) Siberian sturgeon
(*A. baerii*)

(b) White sturgeon
(*A. transmontanus*)

Figure 3: Alien sturgeon species in the Danube River basin (all fotos except (d): Author; (d) Matthias Maier, Austria)

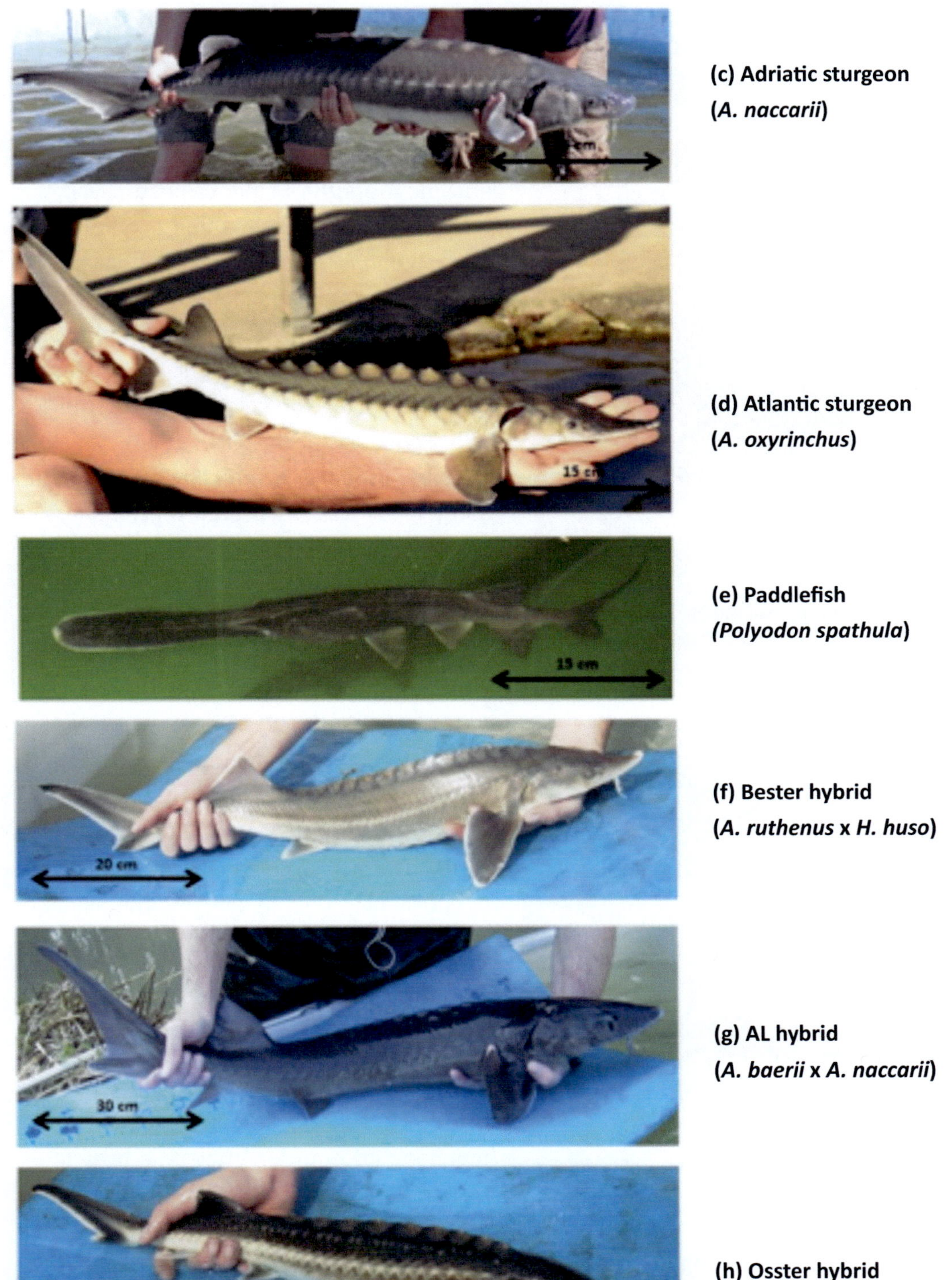

Figure 3 (con.): Alien sturgeon species in the Danube River basin (all fotos except (d): Author; (d) Matthias Maier, Austria)

4.   Economic and cultural value

Figure 4: Beluga sturgeon head sculpture, found near the Iron Gates, dating back about 8500 years. (as proposed by Ivana RADOVANOVIC, 1997) photo by Dragoslav Srejović in 1972

Figure 5: Historic painting of a sturgeon fence on the Romanian Danube by Ludwig Ermini. (retrieved from the archive of Dr. Gertud Haidvogl, Institute of Hydrobiology and Aquatic Ecosystem Management, Vienna, Austria)

Because of their size and the easy catch, sturgeon meat andthe highly valuead caviar was an important food source and economic factor for humans. Besides meat and caviar, the swimbladder, skin and bones were also utilized for producing glue and leather (MOHR, 1952). Sturgeon fishing might have been an important factor for colonization of the Danube River basin (BALON, 1968). A stone sculpture, probably resembling a beluga sturgeon (Fig. 4) was discovered on the shore of the Danube at Lepenski Vir in the Iron Gate area. The sculpture was crafted in the Mesolithic, 8,500 years BC. It is believed, that the annual migration of sturgeon was a symbol of fertility to the inhabitants of the Danube valley (RADOVANOVIC, 1997).

Written records about sturgeon fishing along the Danube date back at least 3500 years BC (HOCHLEITHNER, 2004; KIRSCHBAUM, 2010). The fish were caught with harpoons, nets or fences (Fig. 5) covering the whole or parts of the river. Later, sturgeons were intensively fished by the ancient Romans. Legions, stationed along the Danube, depended on sturgeons as food source (BALON, 1968). As the numbers of fences grew in the fifth century, fishing regulations were implemented in order to avoid a total interruption of the sturgeon migration (BALON, 1968). The fences were restricted in their numbers and their size, allowing only half river - width coverage (SPINDLER, 1997). In 1053 the army of the German Emperor Heinrich the Third may have starved if soldiers would not have caught 50 giant beluga sturgeons (BALON, 1968). A document from 1230 states King Bèla IV. had presented 200 beluga sturgeons annually to the monastery Heiligenkreuz as a gift (BALON, 1968). BUSNITA (1967) estimated catch numbers for the Danube Delta alone have reached around 2,000 tonnes in good years. Sturgeon butcher became a common profession (Fig. 6). On some days up to 450 sturgeons were sold on the Viennese fish market, their weight exceeded

50 tonnes (SPINDLER, 1997). Around the year 1500 a fence, blocking the whole width of the Danube was built near Budapest. Due to protests and politics the fence was destroyed and rebuilt several times (BALON, 1968). Beluga sturgeon fishing was such an attraction that the Viennese royal court visited the fences regularly (BALON, 1968). It is said that canons were used to shoo the sturgeons into the fence traps (BALON, 1968; SPINDLER, 1997). Salted meat was provided to Vienna, and exported Praha, Munich and Paris (BALON, 1968). Because of over exploitation of the stocks, the numbers caught began to drop in the 16th century and the sturgeons, once a food source for the general public, became reserved for the privileged (BALON, 1968). The fishing grounds near Vah became property of the royal court (KHIN, 1957; BARTOSIEWICZ & BONSALL, 2008). The large sturgeon species nearly vanished from Austrian waters in the first half of the 19th century, leaving only the sterlet in considerable numbers (HECKEL & KNER, 1857). At the beginning of the 20th century, sturgeons were already very rare on the Viennese fish market, single specimens being obtained from the middle and lower Danube stretches. Only the sterlet was still common but most fish were also imported from the middle and lower parts of the Danube (KRISCH, 1900), where it was still commercially fished until the beginning of the 21st century (HOLCIK, 1989; GUTI, 2006; GUTI, 2008). Management restrictions for the whole Danube basin were formulated in 1895 at a conference held in Vienna (ANTIPA, 1905), nevertheless overfishing continued and with increasing habitat loss due to river fragmentations stocks continued to dwindle. After the construction of the Iron Gate Dams in the second half of the 20th century, fishery was concentrated in the

Figure 6: Sturgeon butchering in Hamburg in the 19th century. (cited from MOHR, 1952)

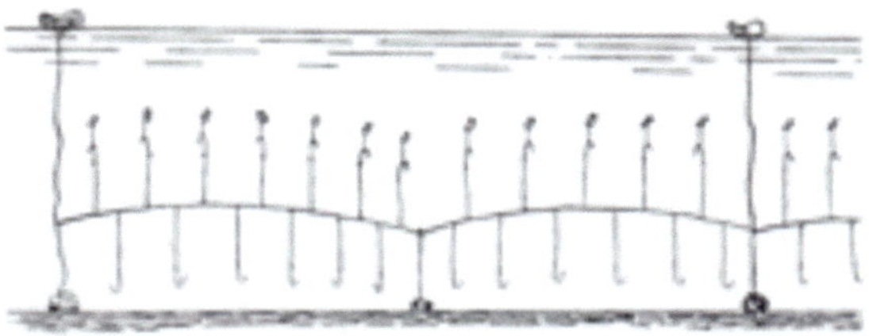

Figure 7: Schematic of a longline used to catch sturgeons. (by M. HOCHLEITHNER)

Figure 8: King Sigismund in a boat carried by two beluga sturgeons in the 15th century. (cited from WINDECK, 1445 - 1450)

lower parts of the Danube. Fishermen were using nets and longlines (Fig. 7) to catch sturgeons As stocks continued to decrease, Romania banned sturgeon fishing in 2006 for 10 years, with Bulgaria following with a ban shortly thereafter (WSCS, 2011).

5. Historical records, recent catches and restoration potential (by river basin)

5.1. Inn and Salzach rivers

5.1.1. Historic records

Reports on catches of sterlets range cover the years from 1890 to 1952, the most upstream ones being two catches in 1907 and 1927 near Rosenheim (MAIER, 1908; MARGREITER, 1927; JUNGWIRTH et al., 1989). STREIBL reported three catches near Marktl and Malching in 1890, 1901 and 1902 (STREIBL; REINARTZ, 2008). The last reports were from the area around Schärding between 1945 and 1952, where an unknown number of catches was stated by

Figure 9: Painting of the sturgeon caught 1746 near Ering. (cited from BROD, 1980)

FISCHER (1952). A sturgeon catch in 1746 near Ering, claimed to be a beluga sturgeon by some authors by noted as "dick" (Russian sturgeon or ship sturgeon) by others (BROD, 1980) can be identified as Russian sturgeon according to the painting (**Fig. 9**). Around the year 1800 there was a record for a specimen of Beluga sturgeon caught near Reichersberg (FREUDLSPERGER, 1936), while KERSCHNER (1956) dates the same catch to around 1880. The fish was preserved and kept in the monastery Reichersberg. Unfortunately, it was destroyed later, so that the systematic status of this specimen remains unclear (SCHMALL & RATSCHAN, 2011).

In the Salzach River the sterlet is claimed by some authors to have been rare (AIGNER & ZETTER, 1859; PEYRER, 1874), while others say it occurred frequently (FREUDLSPERGER, 1936). According to SCHMALL & RATSCHAN (2011) the species was common until the 1800s and was rarely caught in later years. There is one reported catch near Laufen in the midst of the 19th century (HECKEL, 1854; SIEBOLD, 1863). The last known catch at the end of the 1850s by Josef Aigner was kept alive in the castle Mirabell (AIGNER & ZETTER, 1859; ZETTER, 1862; SCHMALL & RATSCHAN, 2011). The catch of a Russian sturgeon, pictured in the castle Hellbrunn near Salzburg, was thought by some authors to be of Salzach origin (ZAUNER, 1997). The diocese Passau, however, had fishing rights not only along the Salzach but also at the lower Inn and the Danube, therefore the origin of this fish cannot be determined accurately (SCHMALL & RAT-

SCHAN, 2011). There is also a portrayal of a beluga sturgeon (**Fig. 10**) in the castle Hellbrunn, with its subtext stating that the fish was caught near Tittmonig in 1617 having a weight of 237 pounds and a length of 3 meters (HOCHLEITHNER, 2004).

Fig. 10: Replica of the beluga sturgeon painting in the Hellbrunn Castle near Salzburg (Foto taken by Author)

5.1.2.   Stocking and recent catches

There is no information available on stocking of sturgeons in the Inn and the Salzach. However, some annecdotal reports claim occasional sturgeon catches in the impoundment Stammham, the mouth of the Alz River (Inn) and the mouth of the Alz- Channel (Salzach) (SCHMALL & RAT-SCHAN, 2011). There is evidence on the catch of a Siberian sturgeon in the mouth of the river Alz in 2003 (Fig. A3) and one Russian sturgeon, one Siberian sturgeon and a specimen of unknown species (probably also a Siberian sturgeon as the catches were reported as one "sturgeon" and two "sterlets") in the impoundment of Stammham in 2004 (Fig. A2 & A4). A white sturgeon (Fig. A1), also reported as "sterlet", was caught in the lower Salzach around 2002 (FRIEDRICH, 2009).

5.1.3.   Potential for population recovery

The lower Inn has five power plants along the Austrian - German border and is therefore highly fragmented. With the average length of the impoundments of approximately 14 km it is very unlikely that a viable, self- sustaining population of sterlets can be supported. In the Salzach stretch, downstream of Laufen (where the natural limit of sturgeon distribution is due to a change in hydromorphology) 44 km of free flowing river are still connected to a 13 km long section of the Inn. All recent catches of sturgeons were either in this section or in the impoundment of the Inn at the power plant directly upstream of this stretch. These findings lead

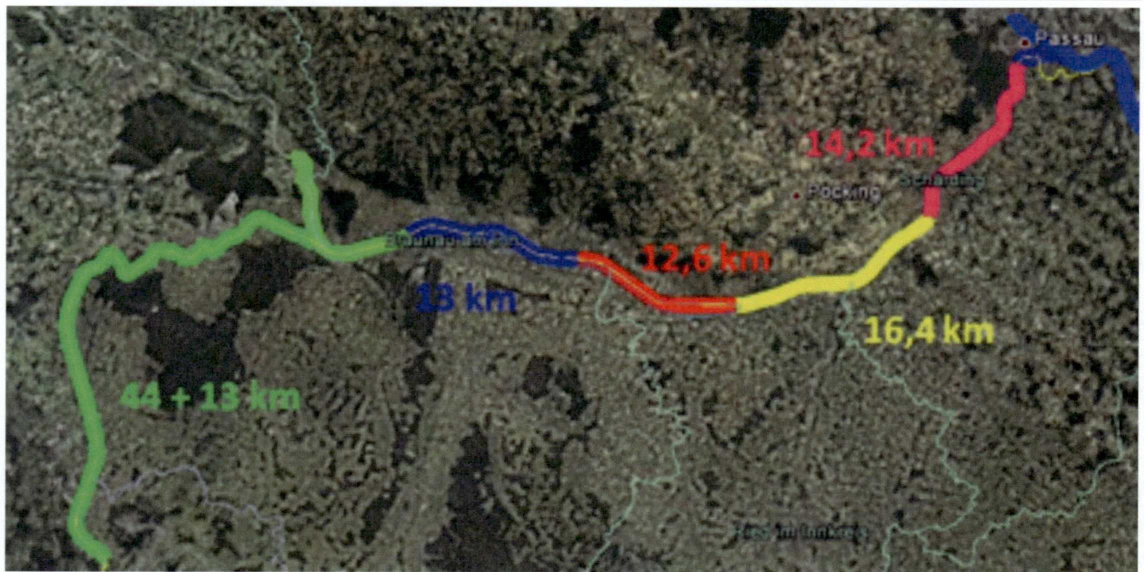

Figure 11: Fragmentation of the Salzach and Inn river systems, identified by different colours for each of the individual impoundments to visualize the migratory restrictions within the river caused by man-made barriers between which the species are trapped. Neighboring stretches with similar colours (f. e. blue) are connected. These fragmentations are super-imposed on Google Earth fotos. Numbers next to each coloured section indicate length of the impoundment in km. (Google/Geoimage Austria)

to the suggestion that this is the only remaining section in the Austrian Inn – Salzach system with certain potential for re-introducing the sterlet. How much habitat for sturgeons is really available in this area has yet to be determined as due to changes in hydromorphology, like straightening, the Salzach River downstream of Laufen developed from a cyprinid river type to a salmonid river type ("rhithralization effect").

5.2.   Danube River

5.2.1.   Historic records

The sterlet was stated as frequent in the Lower Austrian Danube in the 19th century (FITZINGER & HECKEL, 1836). Studying the fish collection of the Museum of Natural History of Vienna brought up 14 specimens caught in or near Vienna between 1831 and 1960. The size of these fish ranged between ~30 cm to ~70 cm total length. Also an albino specimen was caught in 1877.   While some authors state that the species was common in Upper Austria (FITZINGER & HECKEL, 1836; HECKEL & KNER, 1857; BREHM, 1910), others claim it was rare (LORI, 1871; KERSCHNER, 1956). KERSCHNER (1956) reports a catch sold on the fish market in Linz in 1905. Interestingly, there is much more information available for the Bavarian Danube.   Between 1457 and 1905 a total of 13 catches were documented up to Ulm (Tab. 1). The biggest specimens of these were caught near Deggendorf in 1886 with a total length of around 100 cm (STREIBL; REINARTZ, 2008). JAECKEL (1864) reports 3 to 4 catches per year near Passau in the midst of the 19th century and KINZELBACH (1994) states a temporarily high abundance of sterlets in the area of Ulm in the Middle Ages. There are also reports of one specimen from the Isar

Table 1: Historic catches in the Austrian and Bavarian Danube and two Bavarian tributaries (Isar and Lech river systems); TL – total length; Collection ID in the respective reference or Museum code number (compiled from various sources)

Location	Species	Size (cm TL)	Weight (kg)	Year	Note	Source (Collection ID)
Upper Austria	Sterlet			1905		KERSCHNER, 1956
Vilshofen	Beluga sturgeon		81	1605		WACHA, 1956
Regensburg	Ship sturgeon ?		6			FITZINGER & HECKEL, 1836
Steppberg	Sterlet			1673		FITZINGER & HECKEL, 1836
Regensburg	Russian sturgeon		18	1679		FITZINGER & HECKEL, 1836
Straubing	Beluga sturgeon	over 5 feet		1692		FITZINGER & HECKEL, 1836
Regensburg	Sterlet					SIEBOLD, 1863
Regensburg	Sterlet					SIEBOLD, 1863
Bogen	Sterlet					SIEBOLD, 1863
Landshut (Isar)	Sterlet		2,5	1861		SIEBOLD, 1863
Ulm	Sterlet	55	1,25	1822		SIEBOLD, 1863
Deggendorf	Sterlet	100	10	1886		STREIBL; REINARTZ 2008
Vilshofen	Sterlet	56		1887		STREIBL; REINARTZ 2008
Vilshofen	Sterlet	45		1881		STREIBL; REINARTZ 2008
Passau	Sterlet			1864	3 - 4 ind./year	JAECKEL, 1864
Thierhaupten (Lech)	Sterlet			1786		JAECKEL, 1864
Donauwörth	Russian sturgeon ?		82,5	1457		JAECKEL, 1864
Passau	unknown			1852		JAECKEL, 1866
Vienna	Ship sturgeon	~160	~20	1936		ZAUNER 1997; ÖFW 1936
Bad Deutsch- Altenburg	Sterlet	~50		1959		Collection NHM (63055)
Vienna	Sterlet	~35		1884		Collection NHM (63108)
Haslau	Sterlet	~35		1831		Collection NHM (63126)
Vienna	Sterlet	~35		1877	Albino	Collection NHM (63119)
Vienna	Sterlet	~35		1877	five specimens	Collection NHM (63123/63124)
Vienna	Sterlet	~40		1877		Collection NHM (63129)
Vienna, Reichsbrücke	Sterlet	~60		1958		Collection NHM (86430)
Hainburg	Sterlet	~70		1960		Collection NHM (76776)
Vienna	Sterlet	~70		1958		Collection NHM (90197)
Vienna, Danube Channel	unknown	~160		1886		HUGO, 1886

River near Landshut in 1861 (SIEBOLD, 1863) and the Lech near Thierhaupten (JAECKEL, 1864). MOHR (1952) stated that a dedicated fishery for sterlet in Austria does not pay off.

The ship sturgeon was said to migrate up to Komorn and to rarely enter the Austrian Danube (FITZINGER & HECKEL, 1836). In 1936 one specimen was caught in Vienna with a length of around 160 cm TL (**Fig. 12**). The original text describes it as Russian sturgeon (ÖSTERREICHS FISCHEREIWIRTSCHAFT, 1936) whereas ZAUNER (1997) correctly identifies this specimen as ship sturgeon. FITZINGER & HECKEL (1836) also wrote about a catch near Regensburg, the identity of which was doubted by others (SIEBOLD, 1863), as the authors didn`t determine the

Figure12: Ship sturgeon caught 1936 in Vienna. (cited from ÖSTERREICHS FISCHEREIWIRTSCHAFT, 1936)

species of the fish themselves. JAECKEL (1864), with an example of the original catch report and a picture available, stated that according to the picture the fish was most likely *Acipenser nudiventris*.

According to FITZINGER & HECKEL (1836), the Russian sturgeon also entered the Austrian Danube on rare occasions only. The authors distinguished two forms of the Russian sturgeon, *Acipenser gueldenstaedtii*, migrating up to Komorn and *Acipenser schypa* migrating up to Bratislava. This could be an indication for a separate freshwater form of this species. In the same book, a reported catch of a 36 - pound fish near Regensburg in 1679, was described (FITZINGER & HECKEL, 1836). Another catch near Donauwörth in 1457, with a weight of 165 pounds was described as a Russian sturgeon (JAECKEL, 1864). However there is no indication how the author had identified the species.

Regarding the stellate sturgeon there are only two reports from the upper Danube basin. FITZINGER & HECKEL (1836) and MOHR (1952) noted its migration up to Komorn and the species very rarely entering Austrian waters. SIEBOLD (1863) also stated that it rarely occurred in the Isar in Bavaria.

The beluga or great sturgeon was once very abundant in Austrian waters. Due to over-fishing it became very rare as early as the beginning 19th century, migrating mostly up to Bratislava (FITZINGER & HECKEL, 1836). The same situation also seems to have been very likely for the Russian, ship, and stellate sturgeons. One specimen of the Russian sturgeon with over 150 cm total length was caught in 1692 near Straubing (FITZINGER & HECKEL, 1836). GAMLITSCHEK (1897) reports beluga catches "with great success" near Tulln. In 1605, a specimen with 81 kg was caught near Vilshofen in Bavaria (WACHA, 1856). Another catch was recorded by the same author near Ulm in 1822, however, itwas actually a sterlet (SIEBOLD, 1863).

In 1886 a sturgeon was caught in the Danube Channel in Vienna, with a total length of around 160 cm TL (HUGO, 1886). Unfortunately the exact species remains unclear. A second unspecified individual was caught in 1852 near Passau (JAECKEL, 1866).

Figure 13: Traditional beach seine fishery for sturgeon, probably located on the Danube. (cited from OÖLFV, 1997)

5.2.2. Stocking and recent catches

Although there are various rumours regarding stocking of sterlet and other sturgeon species in the Austrian section of the Danube, our survey gave very few reliable data. SPINDLER (1994) mentions a stocking activity of beluga sturgeon near Linz in 1996. Using a picture of the stocking action (Fig. A27), received by the Upper Austrian Fishing Authorities (Oberösterreichischer Landesfischereiverband), the stocked fish were identified by this author as Russian sturgeon. There is a report about a second stocking of Russian sturgeon in the "Donau - Auen National Park" in 2005 (HOCHLEITHNER, pers. comm.; ÖSTERREICHISCHER FISCHEREIVERBAND, 2005). These activities can be seen mostly as some sort of public - relation actions, as the establishment of a population is highly unlikely. Such measures have to be met with high scepticism, as through hybridization and competition the stocked fish might pose a threat to the remaining natural (native) sterlet populations. There has been ongoing stocking of sterlets in quite some river stretches, however the best information available for the area originates from around and downstream of Vienna. Between 2002 and 2005 an intensive program, initiated by the Wiener Fischereiausschuss, took place upstream and downstream of the Freudenau power plant with fish ranging from 35 - 40 cm. All stocked sterlet in the database were obtained from hatcheries which, to the knowledge of this author, use the Danube stock for reproduction.

In the impoundment of the Aschach power plant (for a short description see Annex III) and especially at its head directly below the Jochenstein power station, sterlets of all age classes are caught since the 1950ies (ZAUNER, pers. comm.; WAIDBACHER, pers. comm.). The species was and still is rather common and many fishermen rarely record their catches. Even

Table 2: Stocking of sturgeons into different locations of the Austrian Danube during the last 30 years. (quantity = numbers released). The respective figures of the species are given by the figure number in the Annex 1

Location	Species	Size (cm TL)	Quantity	Year	Note	Source
Linz	Russian sturgeon	70 - 80	?	1996	Fig. A27	OÖ LFV, 1996
Wachau	Sterlet	15 - 25	~700	1994		KIWEK, 1995
above pp Freudenau	Sterlet	?	?	?		HOCHLEITHNER, pers. comm.
above pp Freudenau	Sterlet	35 - 40	~7000	2002 -2005	Fig. A38	HP of WFA
below pp Freudenau	Sterlet	35 - 40	~3000	2002 -2006	Fig. A38	HP of WFA
Albern harbour	Sterlet	30 - 40	5	2010	Fig. A43	SCHEIBLECHNER, pers. comm.
Donau- Auen National Park	Russian sturgeon	8-12	300	2005	Fig. A41 & A42	ÖSTERREICHS FISCHEREI, 2005; HOCHLEITHNER, pers. comm.
Donau- Auen National Park	Sterlet	15 - 25	~1000	2001	Fig. A40	HOCHLEITHNER, pers. comm.
Regelsbrunn	Sterlet	15 - 25	~1000	1994	Fig. A39	HOCHLEITHNER, pers. comm.
near mouth of March	Sterlet	15 - 25	~700	1994		KIWEK, 1995

with just a portion of the actual catch numbers in the database, there are 22 sterlet catches recorded between 2002 and 2011 (Tab. 3). Most are caught directly downstream of the Jochenstein power plant, using nets. The furthest downstream record was near Obermühl. To this point there are no indications of the species using the slow - flowing impoundment area upstream of the Aschach power station. In comparison to other fish species the sturgeons cope

*Tab. 3 (continues on next page): Sturgeon catches in the Austrian Danube during the last 30 years, listed for different impoudments, including Aschach, Ottenscheinm Abwinden-Asten, Wallsee-Mitterkirchen, Ybbs-Persenbeug, Altenwörth, Greifenstein, Freudenau, Gabcikovo. *HP = internet homepage of the respective reference as listed at the*

Impoundment	Location	Species	Size (cm TL)	Weight (g)	Year	Note	Source
Aschach	below pp Jochenstein	Sterlet	~50		2002	Fig. A5	ZAUNER, pers. comm.
Aschach	below pp Jochenstein	Sterlet	~35		2002	Fig. A5 & A6	ZAUNER, pers. comm.
Aschach	below pp Jochenstein	Sterlet	~60		2002	Fig. A7	ZAUNER, pers. comm.
Aschach	Schlögen	Paddlefish	115	8700	2003	Fig. A21	HP of LFVOOE*
Aschach	Niederranna	Sterlet	76	1335	2005	Fig. A20	HP of HOFKIRCHEN*
Aschach	below pp Jochenstein	Sterlet	~85		2006	Fig. A8	REINARTZ, 2008
Aschach	below pp Jochenstein	Sterlet			2006/2007	six specimens	REINARTZ, 2008
Aschach	below pp Jochenstein	Siberian sturgeon	~75		2006/2007	Fig. A9	REINARTZ, 2008
Aschach	below pp Jochenstein	Hybrid Sibster			2006/2007	four specimens	REINARTZ, 2008
Aschach	below pp Jochenstein	Sterlet	~45		2007	Fig. A10	ZAUNER, pers. comm.
Aschach	below pp Jochenstein	Sterlet	50		2008	Fig. A11	ZAUNER, pers. comm.
Aschach	below pp Jochenstein	Sterlet	~60		2008	Fig. A12	ZAUNER, pers. comm.
Aschach	Obermühl	unknown	60		2010		HP of FISCHERFORUM*
Aschach	below pp Jochenstein	Sterlet	~45		2011	Fig. A18	ZAUNER, pers. comm.
Aschach	below pp Jochenstein	Sterlet	54		2011	Fig. A14	ZAUNER, pers. comm.
Aschach	below pp Jochenstein	Sterlet	49		2011	Fig. A15	ZAUNER, pers. comm.
Aschach	below pp Jochenstein	Sterlet	46		2011	Fig. A16	ZAUNER, pers. comm.
Aschach	below pp Jochenstein	Sterlet	49		2011	Fig. A17	ZAUNER, pers. comm.
Aschach	below pp Jochenstein	Sterlet	35		2011	stocked in Schwarzer Laber Fig. A13	ZAUNER, pers. comm.
Aschach	below pp Jochenstein	Sterlet	35		2011	stocked in Schwarzer Laber	ZAUNER, pers. comm.
Aschach	mouth of Dandlbach	Sterlet	~30		2011	Fig. A19	ZAUNER, pers. comm.
Aschach	Schlögen	unknown			2011		ZAUNER, pers. comm.
Ottensheim	Alkoven	Paddlefish	120	8500	1993		HOLCIK, 2006
Abwinden - Asten	Linz	Hybrid	~115			possibly *A. naccarii* x *H. huso* Fig. A23	ANONYMOUS, 2005
Abwinden - Asten	Linz	Sterlet	~45		1984	Fig. A22	WIESMAYER, pers. comm.
Abwinden - Asten	Linz	Siberian sturgeon	~75		1999	Fig. A25	WIESMAYER, pers. comm.
Abwinden - Asten	Linz	unknown	~60		2011		WIESMAYER, pers. comm.
Abwinden - Asten	Abwinden	Hybrid	~110		2003	possibly *A. naccarii* x *H. huso* Fig. A24	HP of HECHTSPRUNG*
Wallsee - Mitterkirchen	below pp Abwinden	unknown		5000 -7000	2001-2003		HP of FISCHERFORUM*

Impoundment	Location	Species	Size (cm TL)	Weight (g)	Year	Note	Source
Ybbs - Persenbeug	below pp Wallsee	Sterlet	40		1997		ANGERER, pers. comm.
Ybbs - Persenbeug	below pp Wallsee	Sterlet	50		1997	Fig. A26	ANGERER, pers. comm.
Altenwörth	Hollenburg	unknown	38		2002		HP of DONAUFISCHER*
Altenwörth	Dürnstein	Russian sturgeon	100	4000	2003	Fig. A28	
Altenwörth	below pp Melk	Sterlet	91		2007	Fig. A29	TROST, pers. comm.
Altenwörth	below pp Melk	unknown	~95		2009		HP of ANGELFORUM*
Altenwörth	Emmersdorf	unknown			2011		FÜRNWEGER, pers. comm.
Greifenstein	below pp Altenwörth	Sterlet	54		1980		KOVARIK, pers. comm.
Greifenstein	below pp Altenwörth	Hybrid	~115		2002	possibly *A. gueldenstaedtii* x *H. huso* Fig. A32	JUNGWIRTH, pers. comm.
Greifenstein	below pp Altenwörth	Sterlet	~90		2003	Fig. A35	ELSBACHER, pers. comm.
Greifenstein	below pp Altenwörth	unknown	~90		2005		ELSBACHER, pers. comm.
Greifenstein	below pp Altenwörth	Siberian sturgeon	~110		2005	Fig. A33	JUNGWIRTH, pers. comm.
Greifenstein	Zwentendorf	unknown			2010		KOVARIK, pers. comm.
Greifenstein	below pp Altenwörth	Siberian sturgeon	96		2011	Fig. A34	JUNGWIRTH, pers. comm.
Freudenau	Langenzersdorf	Sterlet	25 - 35		1986		KROMP, pers. comm.
Freudenau	Langenzersdorf	Sterlet	25 - 35		1988		KROMP, pers. comm.
Freudenau	Langenzersdorf	Sterlet	25 - 35		1990		KROMP, pers. comm.
Freudenau	Klosterneuburger Durchstich	Sterlet	~100		1993		WAIDBACHER, pers. comm.
Freudenau	Klosterneuburger Durchstich	Sterlet			1999	two specimens	WAIDBACHER & STRAIF, 2005
Freudenau	Klosterneuburger Durchstich	Sterlet	~60		1990-1994		DELLINGER, pers. comm.
Gabcikovo	below pp Freudenau	Sterlet	70		1999	Fig. A36	EBERSTALLER, et al., 2001
Gabcikovo	below pp Freudenau	Sterlet	56		1999		EBERSTALLER, et al., 2001
Gabcikovo	below pp Freudenau	Sterlet	74		2000		EBERSTALLER, et al., 2001
Gabcikovo	below pp Freudenau	Sterlet	67		2000	two specimens	EBERSTALLER, et al., 2001
Gabcikovo	below pp Freudenau	Sterlet	70		2000		EBERSTALLER, et al., 2001
Gabcikovo	below pp Freudenau	Sterlet	68		2000	two specimens	EBERSTALLER, et al., 2001
Gabcikovo	Freudenau harbour	unknown	72		2009/2010		HP of WFA*
Gabcikovo	Danube- Oder channel	unknown	40 - 50		2009/2010	eight specimens	HP of WFA*
Gabcikovo	Freudenau harbour	unknown			2011	two specimens	PETROUSCHEK, pers. comm.
Gabcikovo	Freudenau harbour	unknown			2011		ZENS, pers. comm.
Gabcikovo	Freudenau	Sterlet	75	3152	2011	Fig. A37	SCHLAPPAL, pers. comm.
Gabcikovo	Cunovo	Paddlefish	114	8200			HOLCIK, 2006
unknown		Siberian sturgeon	~115			Fig. A30	
unknown		Siberian sturgeon	~130		2007/2008	Fig. A31	JUNGWIRTH, pers. comm.

well with the stress of being entangled in a net and show no injuries or odd behaviour (ZAUNER, pers. comm.). The caught fish in the database range from around 30 cm TL (age 2 years) to around 85 cm TL (at least 10 years old). In 2011 two tagged sterlets were caught (ZAUNER, pers. comm.) (Fig. A13), which were stocked in the Bavarian river Schwarze Laber (HIRMER, 2011). These specimens migrated around 150 km downstream and passed several power plants. A disturbing discovery was made in 2006/07, when several hybrids and a mature Siberian sturgeon were caught in this area (REINARTZ, 2008). These hybrids between male Siberian sturgeon and female sterlet give the first indication for hybridization between native and alien sturgeon species in a river system and of the Siberian sturgeon maturing to a stage potentially permitting natural spawning outside of its natural range (LUDWIG et al., 2009).

Other species and hybrids have been caught in various sections of the Upper Austrian Danube, like paddlefish near Schlögen (LFVOOE, 2003) and Alkoven (HOLCIK, 2006), Siberian sturgeon (WIESMAYER, pers. comm.) and hybrids (ANONYMOUS, 2005) near Linz and fish of unknown species near Linz (WIESMAYER, pers. comm.) and Abwinden.

A sterlet, caught in 1984 near Linz (WIESMAYER, pers. comm.) is of interest, as it might have migrated downstream from the Jochenstein population or was a remnant in this Danube section (Fig. A22). Below the Wallsee - Mitterkirchen power plant two Sterlets have been caught in 1997, with the fisherman observing another two specimens feeding in a backwater (ANGERER, pers. comm.) (Fig. A26).

There are no reported catches for the Melk impoundment and very few for one of the two last free- flowing sections in Austria, the Wachau, on its downstream end dammed by the Altenwörth power station. Three catches of unknown species could be found (Tab. 3) and one Russian sturgeon (Fig. A28), reported as beluga sturgeon by the fisherman. A very large sterlet was caught at an unknown point (Fig. A29), probably a relic of a former population.

Downstream of the Altenwörth power station a sterlet with 54 cm was caught in 1980 (KOVARIK, pers. comm.). A very big specimen with around 90 cm (ELSBACHER, pers. comm.) and at least 10 years age was caught in 2003 (Fig. A35). Both specimens are probably either migrants from other stretches or relics of a former population. Also various exotic species and hybrids have been caught in this stretch, especially after the 2002 flood. Most of them identified as big Siberian sturgeons, all of which in very good condition (Fig. A33 & A34 and probably A30 & A31).

Before construction of the Freudenau power station, juvenile sterlets have been caught near Langenzersdorf and Klosterneuburg between 1986 and 1990 (KROMP, pers. comm.). Later catches were adult fish ranging from 60 cm to 100 cm (WAIDBACHER, pers. comm.; DELLINGER, pers. comm.; WAIDBACHER et al., 2006), with four fish being recorded. Although intensive stocking was carried out between 2002 and 2005, no catches in this area could be recorded in the following years.

In 1999 and 2000 EBERSTALLER et al. (2001) could catch eight specimens of sterlet directly below the Freudenau dam during a survey of the newly built fish ladder (Fig. A36). These fish, caught with nets, could be wild fish or originate from the stocking of 1994. After the stocking by the Wiener Fischereiauschuss 2002 - 2005 various sturgeon catches occurred in this area. 14 could be recorded in the database of the author, but probably many more occurred and were not published or registered. Likely most of the caught fish are sterlets (for example Fig. A36), but without pictures or exact descriptions identification is nearly impossible. There are also records regarding sightings of fish from the stocking, with over 1 m total length in 2003 (EDER, 2003). However, they cannot belong to the stocked fish from this period due to their size, the size of the stocked fish (Tab. 2) and the average growth rate of sterlets. It remains unclear if these fish are wild remnants, earlier stocked fish or of another sturgeon species released in the Danube. From the Hungarian section of this stretch there are no catch records of sterlets (GUTI, pers. comm.), while HOLCIK (1989) states some catches of sterlets in this section in the 1980ies and also the catch of a paddlefish in 2006 (HOLCIK, 2006). In the Slovakian section sterlets are caught regularly as broodstock, recently also with Siberian sturgeons as by-catch (PEKARIK, pers. comm.).

During printing of this publication another Siberian sturgeon and two Russian sturgeons (Fig. A51) were caught next to some sterlets below the powerplant Jochenstein (ZAUNER, pers. comm.).

5.2.3. Potential for population recovery

According to various scientists (ZAUNER, pers. comm.; WAIDBACHER, pers. comm.; REINARTZ, pers. comm.; FRIEDRICH, 2009) the Jochenstein population is the last remaining, self-sustaining population of the sterlet in the Austrian Danube. (Fig. 14) This population has to be protected and supported by adequate measures to avoid habitat degradations and other negative influences such as contact with alien sturgeon species resulting in hybridization. The increasing number of alien sturgeon species within natural water bodies is of high concern as it poses a serious threat to the small and dwindling native populations and can affect restocking plans for native species in all catchments. As we do not know the exact status and size of the Jochenstein population, nor the actual use of locations and the characteristics of their key habitats

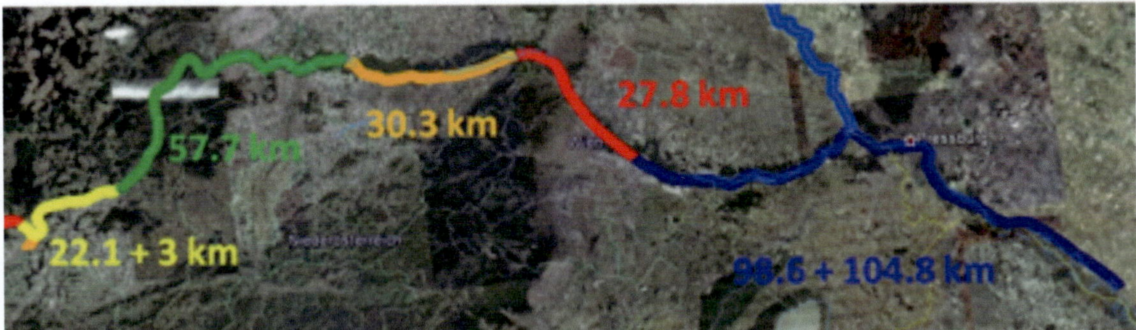

Figure 14: Fragmentation of the Danube system (Top: Upper Austria; bottom Lower Austria), identified by different colours for each of the individual impoundments to visualize the migratory restrictions within the river caused by man-made barriers between which the species are trapped. Neighboring stretches with similar colours (f. e. blue) are connected. These fragmentations are super-imposed on Google Earth fotos. Numbers next to the coloured section indicate length of the impoundment in km. (Google/Geoimage Austria)

within this section (feeding, wintering, spawning and nursing areas), further research is urgently needed to close this knowledge gap. Once the respective habitats are identified and sufficiently described, adequate additional habitats can be searched for or reconstructed in other river stretches.

Other stretches of the Upper Austrian Danube are partially much shorter than the one described above and differ in their hydromorphological characteristics. The connected tailwater sections of the large tributaries Inn, Traun and Enns are only 3 - 4 km long. If the catches of sterlets near Linz in the 1980s and at Mitterkirchen in the 1990s were from native sterlet stocks, these populations have most likely become extinct today. It might be that with ongoing revitalization and habitat improvements, adequate sterlet habitats may become available again. If so, restocking may become an option and should be done with juveniles hatched from the Jochenstein brood stock.

The situation on most Lower Austrian Danube stretches is similar to those of the Upper Austrian waters. The Melk impoundment is very short, with only a very small section of the Ybbs connected to it. In the free- flowing section through the Wachau valley only very few catches have been recorded. Nevertheless it has to be seen as a priority area due to its status and characteristics. It might be possible to find adequate sterlet habitats in the future, especially with ongoing habitat improvements and revitalization programmes, mostly through the LIFE+ projects of the European Union.

The high number of sturgeon catches below the Altenwörth power plant may be the result of the 2002 flood. At least the feeding habitat in this stretch seems to be excellent, as all caught fish were in very good condition. A sterlet caught in 1980 and a very large specimen caught in 2003 (the latter unlikely originating from stocking due to its size), both were probably relics of a former native population. Also the reports of 1 - 2 year old sterlets caught in the late 1980s near Klosterneuburg show that at this time spawning must have taken place downstream of the Greifenstein power station which was completed in 1985. This also indicates the former existence of a sterlet population in this area. After the Freudenau power station was constructed (1992 – 1998) the sterlet vanished in this area and even intense stocking did not show any improvement.

The river section with the highest potential for sturgeon restoration/re-introduction in Austria is the free- flowing section between Vienna and Gabcikovo, comprising the Donau – Auen National Park. This section is around 100 km long and is still connected to around 70 km of the March River and about 30 km of the Thaya River, both previously inhabited by sturgeons. This system is the longest open, free- flowing river system in Austria. It offers both epi - and metapotamal characteristics, subsequently higher habitat variability and a greater chance of habitat availability for all life stages of sturgeons. Human impacts on the hydromorphology in the Thaya - March system are comparatively low. Although in the Danube most shorelines have been stabilized with rip-rap, ongoing revitalization activities are taking place, connecting backwaters and improving habitat, especially in the National Park area. Intensive stocking of sterlets took place at the beginning of the 21st century, and catches increased thereafter. Occasional catches and sightings of larger specimens indicated the existence of a modest number of wild fish. Presently, there is no proof of any natural reproduction of the species in this area, suggest-

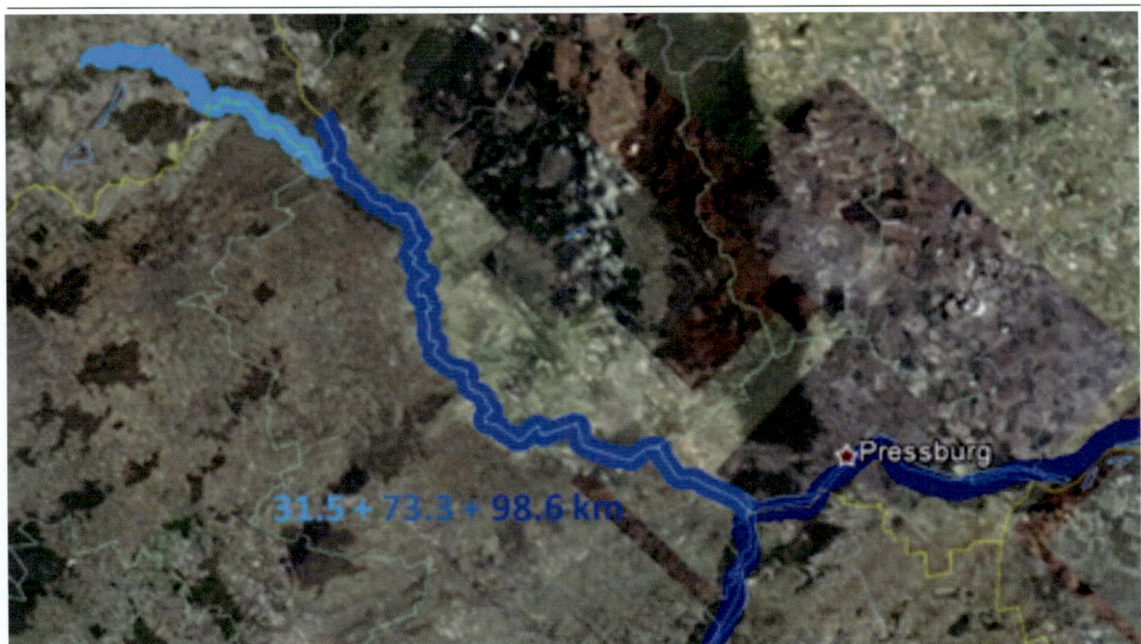

Figure 15: Connected river stretches of the Morava- Danube system, identified by different colours for each of the individual impoundments to visualize the migratory restrictions within the river caused by man-made barriers be-tween which the species are trapped. Neighboring stretches with similar colours (f. e. blue) are connected. These fragmentations are super-imposed on Google Earth fotos. Numbers next to the coloured section indicate length of the impoundment in km. (Google/Geoimage Austria)

ing either the lack of spawning sites, or the low capability of stocked fish to adapt to natural conditions. Another reason could be the insufficient number of wild spawners or mature stocked fish. The increasing density of European catfish (*Silurus glanis*) in this area, might also pose a vigorous predator threat to juveniles. The same can be stated for the large stocks of various allochthonous gobies (*Neogobius melanostomus, Neogobius kessleri, Neogobius gymnotrachelus, Proterohinus marmoratus*) in the entire catchment area which are capable of feeding on larvae and post-larvae (for details see Chapter Predation). Before any actions are taken, it is inevitable to try to determine the reasons for the decline of the sterlet stocks in this area to increase the chances of success for restoration measures. Any efforts in this area should be executed on a multilateral basis between the range states, including Austria, Slovakia, the Czech Republic and Hungary. If in the future the most serious barriers, dams at the Iron Gates and Gabcikovo will be overcome by the construction of adequate fishpasses that can be used by sturgeons, the stretches downstream of Vienna will most likely play an important role as spawning habitats. Constructing further passages through the ten power plants in Austria and thereby greatly improving river section connectivity will possibly be a much more difficult challenge to undertake. In contrast to the anadromous species the reintroduction of the landlocked ship sturgeon in this area might be a more promising option (ZAUNER, 1997; ANONYMOUS, 2004a; ANONYMOUS, 2004b; ANONYMOUS, 2004c; ANONYMOUS, 2004d; FRIEDRICH, 2009). However, as the status of the ship sturgeon is at the brink of extinction and since there is no captive brood stock presently available, such considerations may be of mid- to long-term importance but should be included as potential options in any of the Action Plans under consideration (see also Chapter Stocking and Reintroduction).

5.3.   March and Thaya rivers

5.3.1.   *Historic records*

Various authors stated the occurrence of the sterlet in the March River (HEINRICH, 1856; KRAFT, 1874; REMES, 1902), although no exact locations were given. According to REMES (1902) the species also occurred in the Thaya River. WEEGER (1884) reports of "sturgeon" catches near Rabensburg and Hodonin, which are most likely Russian sturgeons, as the Russian sturgeon is also stated to occur in the March (JEITTELES, 1864; REMES, 1902). The Beluga sturgeon migrated in the past at least up to the Landshut area as there were reports of catches of this fish of up to two meters of size (REMES, 1902; ZBORIL & ABSOLON, 1916). HEINRICH (1856) stated the species occurred in the area around Lundenburg and MAHEN (1927) mentions that it was very rare. For the areas near Rabensburg and Hodonin, beluga sturgeon catches were also reported (WEEGER, 1884).

5.3.2.   *Stocking and recent catches*

There are no records on stocking of sturgeons in the March and Thaya rivers, only a release of sterlets in the Danube near the mouth of the March (see Tab. 2) was carried out in 1994 (KIWEK, 1995). Furthermore, it was not possible to obtain any data about sturgeon catches in the last 30 years except a note of 2- 3 catches annually by commercial fishermen since 1980 (HOLCIK, 1995). According to SPINDLER (1994) it is likely that the sterlet vanished from the March system.

5.3.3.   *Potential for population recovery*
(See Chapter Danube - Potential for population recovery)

5.4.   Other Tributaries

5.4.1.   *Historic records*

Except for the rivers already mentioned, also other large tributaries have been examined, especially the Traun, Enns and Ybbs rivers. Interestingly, the only note obtained for all of these rivers was that sterlet rarely occurred in the mouth of the Enns River (ANONYMOUS, 1884). It is not known whether sturgeons never inhabited the lower parts of these rivers, or if there were just no records. However it seems likely that at least the lowest tailwater stretches close to the Danube were used for spawning in the past. An interesting note is that FRAUENFELD (1871) stocked the Hadersdorfer pond with a few dozens of sterlets from the Danube in the year 1847. Through a flood shortly thereafter most individuals escaped and one was caught in the Vienna River below the Gumpendorfer dam (FRAUENFELD, 1871).

There are no confirmed data available on stocking of sturgeons in these waters, only an oral report from the lower Traun with an unknown number of fish of an unidentified sturgeon species of unknown origin (LAHMER, pers. comm.). The fishing regulations for the lower Traun state sturgeons and beluga sturgeon as protected and this may be an indication of previous releases. Unfortunately, it was not possible to gain further information to confirm any occurrences. It is possible that there is a link to the stocking of sturgeons in the harbour of Linz in 1998 (Tab. 2; Fig. A27). There are two reports of paddlefish in Upper Austrian waters. One specimen (Fig. A44) was found dead in the deterrent screens of a power plant in the Aschach River in 1996 (JAGSCH, 1996; ZAUNER, 1997). A second fish was caught in the Enns River between 2005 und 2010 (JUNGWIRTH, pers. comm.).

5.4.3.   *Potential for population recovery*

Very short tailwater stretches of Traun, Enns and Ybbs, with around three kilometers each, are still connected to the Danube. Also there are many power stations, other barriers and in the Traun even a diverted river section in the lower parts. Other problems are the habitat alterations, the hydromorphological changes and "rhithralization effects" in some parts. Therefore these rivers can be assumed unsuitable for sterlet reintroduction at this point, leaving only their mouths as possible habitat.

5.5.   Drava River

5.5.1.   *Historic records*

Occurrence of the sterlet in the Drava has been proven by various authors (HECKEL & KNER, 1857; GLOWACKI, 1885; MOJSISOVICS, 1897), but there is no indication how far upstream its distribution has been in the past. It is stated to have been a common fish near Warazdin but rarely reached Ptuj. Occasionally it was seen near Maribor (WOSCHITZ, 2006). FITZINGER & HECKEL (1836) also report of the Russian sturgeon, ship sturgeon and beluga sturgeon, HECKEL & KNER (1857) also of the stellate sturgeon. It is unclear if these species only migrated up to Legrad near the mouth of the Mura River, or further upstream to the Ptuj area (WOSCHITZ, 2006). One catch of ship sturgeon in the Mura in 2005 (GUTI, 2006) proofs that this species at least migrates up to the mouth of the Mura. Another specimen was caught further downstream near Heresznye in 1989 (GUTI, 2006). Regarding the available information (HONSIG - ERLENBURG & FRIEDL, 1999; WOSCHITZ, 2006) it remains doubtful if any of the five sturgeon species ever occurred in today`s Austrian Drava.

5.5.2.   *Stocking and recent catches*

The various impoundments of the Drava were stocked from 1982 to 1995 with sterlets. In total around 1000 specimens, mostly ranging from 30 to 50 cm total length and partially from 5 to 20 cm, were stocked in four different impoundments (Annabrücke, Rosegg, Völkermarkt, Lavamünd). The fish were in part wild fish from the Hungarian Danube and some originate from the Danube - stock fish hatched and reared at the hatchery in Szasalombhatta, Hungary (HON-

SIG - ERLENBURG & FRIEDL, 1999). In 2010 300 additional sterlets were stocked by a hydro-power company in the Völkermarkt impoundment. The fish were obtained from a hatchery in Waldschach, Styria (SALZMANN, pers. comm.). Additionally, a fishing club stocked 50 additional sterlets in 2010 in the Annabrücke impoundment (PLEYER, pers. comm.). A questionable stocking may have taken place in 2010, releasing an unknown quantity of Russian sturgeons of unknown origin in the Völkermarkt impoundment (HONSIG - ERLENBURG, pers. comm.).

In the years following the stocking many specimens where caught and reported, especially in the impoundment Annabrücke (69 individuals), mostly in the free- flowing section downstream the power station Ferlach, and the impoundment Völkermarkt (20 individuals)

Table 4: Data collected on recent sturgeon stocking in various impoudments of the Drava River (between 1982 and 2010)

Impoundments	Species	Size (cm TL)	Quantity	Year	Source
Annabrücke	Sterlet	45 - 55	130	1982	HONSIG - ERLENBURG & FRIEDL, 1999
Rosegg	Sterlet	45 - 55	130	1982	HONSIG - ERLENBURG & FRIEDL, 1999
Rosegg	Sterlet	05-ott	150	1983	HONSIG - ERLENBURG & FRIEDL, 1999
Annabrücke	Sterlet	28 - 55	70	1987	HONSIG - ERLENBURG & FRIEDL, 1999
Völkermarkt	Sterlet	28 - 55	300	1987	HONSIG - ERLENBURG & FRIEDL, 1999
Lavamünd	Sterlet	20 - 30	111	1991	HONSIG - ERLENBURG & FRIEDL, 1999
Lavamünd	Sterlet	20 - 30	90	1995	HONSIG - ERLENBURG & FRIEDL, 1999
Völkermarkt	Sterlet	~20	300	2010	SALZMANN, pers. comm.
Völkermarkt	Russian sturgeon	unknown	unknown	2010	HONSIG - ERLENBURG, pers. comm.
Annabrücke	Sterlet	~25	50	2005	PLEYER, pers. comm.

Table 5 (continues on next page): Recorded sturgeon catches in the various impoundments of the Drava River system between 1982 and 2008-. N = number of fish caught; TL = total length; sizes are given for individual fish or when reported in groups the size range is indicated

Impoundment	Species	Size (cm TL)	Weight (kg)	Year	N	Source
Rosegg	Sterlet			1982	1	HONSIG - ERLENBURG & FRIEDL, 1999
Rosegg	Sterlet			1982	1	HONSIG - ERLENBURG & FRIEDL, 1999
Rosegg	Sterlet	35,5	208	1985	1	HONSIG - ERLENBURG & FRIEDL, 1999
Feistritz	Sterlet			1984	2	HONSIG - ERLENBURG & FRIEDL, 1999
Feistritz	unknown	107	8000	2002	2	MACHACEK, pers. comm.
Ferlach	Sterlet	47 - 58		1987	1	HONSIG - ERLENBURG & FRIEDL, 1999
Ferlach	unknown	40	300	1997	1	HONSIG - ERLENBURG & FRIEDL, 1999
Ferlach	unknown			1997	1	HONSIG - ERLENBURG & FRIEDL, 1999
Ferlach	unknown	92	4500	1998	1	HONSIG - ERLENBURG & FRIEDL, 1999
Ferlach	unknown			1998	1	HONSIG - ERLENBURG & FRIEDL, 1999
Ferlach	unknown	55		1998	1	HONSIG - ERLENBURG & FRIEDL, 1999
Ferlach	unknown	86		2000	1	KOGLER, pers. comm.
Ferlach	unknown			2001	2	KOGLER, pers. comm.

Impoundment	Species	Size (cm TL)	Weight (kg)	Year	N	Source
Annabrücke	Sterlet			1983	1	HONSIG - ERLENBURG & FRIEDL, 1999
Annabrücke	Sterlet	35		1983	1	HONSIG - ERLENBURG & FRIEDL, 1999
Annabrücke	Sterlet	50		1984	1	HONSIG - ERLENBURG & FRIEDL, 1999
Annabrücke	Sterlet	50	590	1985	1	HONSIG - ERLENBURG & FRIEDL, 1999
Annabrücke	Sterlet	34	100	1987	1	HONSIG - ERLENBURG & FRIEDL, 1999
Annabrücke	Sterlet	63		1988	1	HONSIG - ERLENBURG & FRIEDL, 1999
Annabrücke	Sterlet			1993	5	HONSIG - ERLENBURG & FRIEDL, 1999; PLEYER, pers. comm.
Annabrücke	unknown			1994	2	HONSIG - ERLENBURG & FRIEDL, 1999; PLEYER, pers. comm.
Annabrücke	unknown			1995	2	HONSIG - ERLENBURG & FRIEDL, 1999; PLEYER, pers. comm.
Annabrücke	unknown	90		1996	1	HONSIG - ERLENBURG & FRIEDL, 1999
Annabrücke	unknown			1996	3	HONSIG - ERLENBURG & FRIEDL, 1999; PLEYER, pers. comm.
Annabrücke	unknown	96		1997	1	HONSIG - ERLENBURG & FRIEDL, 1999
Annabrücke	unknown			1997	3	HONSIG - ERLENBURG & FRIEDL, 1999; PLEYER, pers. comm.
Annabrücke	unknown			1998	2	PLEYER, pers. comm.
Annabrücke	unknown			1999	4	PLEYER, pers. comm.
Annabrücke	unknown			2000	1	PLEYER, pers. comm.
Annabrücke	unknown			2001	5	PLEYER, pers. comm.
Annabrücke	unknown			2002	7	PLEYER, pers. comm.
Annabrücke	unknown			2003	2	PLEYER, pers. comm.
Annabrücke	unknown			2004	4	PLEYER, pers. comm.
Annabrücke	unknown			2005	7	PLEYER, pers. comm.
Annabrücke	unknown			2006	8	PLEYER, pers. comm.
Annabrücke	unknown			2007	3	PLEYER, pers. comm.
Annabrücke	unknown			2008	2	PLEYER, pers. comm.
Völkermarkt	Sterlet	47		1982	1	HONSIG - ERLENBURG & FRIEDL, 1999
Völkermarkt	Sterlet	52		1982	1	HONSIG - ERLENBURG & FRIEDL, 1999
Völkermarkt	Sterlet	30 - 52		1982	5	HONSIG - ERLENBURG & FRIEDL, 1999
Völkermarkt	Sterlet	40 - 60		1983	5	HONSIG - ERLENBURG & FRIEDL, 1999
Völkermarkt	Sterlet	65		1985	1	HONSIG - ERLENBURG & FRIEDL, 1999
Völkermarkt	Sterlet	72	1700	1990	1	HONSIG - ERLENBURG & FRIEDL, 1999
Völkermarkt	Sterlet	52	690	1990	1	HONSIG - ERLENBURG & FRIEDL, 1999
Völkermarkt	Sterlet			1992	1	HONSIG - ERLENBURG & FRIEDL, 1999
Völkermarkt	Sterlet	55		1992	1	HONSIG - ERLENBURG & FRIEDL, 1999
Völkermarkt	unknown	78	2000	1996	1	HONSIG - ERLENBURG & FRIEDL, 1999
Völkermarkt	unknown	70		1997	1	HONSIG - ERLENBURG & FRIEDL, 1999
Völkermarkt	Sterlet	2,5		1998	1	HONSIG - ERLENBURG & FRIEDL, 1999
Schwabegg	Sterlet			1988	1	HONSIG - ERLENBURG & FRIEDL, 1999
Schwabegg	Sterlet			1989	1	HONSIG - ERLENBURG & FRIEDL, 1999
Schwabegg	Sterlet	68		1990	1	HONSIG - ERLENBURG & FRIEDL, 1999
Schwabegg	Sterlet	66	800	1991	1	HONSIG - ERLENBURG & FRIEDL, 1999
Lavamünd	Sterlet	50 - 55		1988	2	HONSIG - ERLENBURG & FRIEDL, 1999
Lavamünd	Sterlet	52,5	500	1993	1	HONSIG - ERLENBURG & FRIEDL, 1999
Lavamünd	unknown	50 - 60		1998	4	HONSIG - ERLENBURG & FRIEDL, 1999
Lavamünd	Sterlet	94	4800	2003	1	HP of FISCHERFORUM* (Fig. A45)

(HONSIG - ERLENBURG & FRIEDL, 1999; PLEYER, pers. comm.). In total 45 specimens were re-ported between 1982 and 1993 and 74 specimens between 1994 and 2008.The biggest speci-men reached 107 cm TL and 8 kg which was being caught in the impoundment Feistritz in 2002 (MACHACEK, pers. comm.). The fish caught before 1993 were most certainly sterlets. Many fish caught later were also probably sterlets, but probably other species have been introduced thereafter. After the stocking, sterlets were also caught in the Slovenian Drava near the border (HONSIG - ERLENBURG & FRIEDL, 1999). Of interest is that at least some specimens were able to pass the barriers of the power plant dams unharmed through the turbines (HONSIG - ER-LENBURG & PETUTSCHNIG, 2002). A stomach examination of a fish captured in 1993 showed that it mostly fed on larvae of caddisflies, belonging to the genus *Hydropsyche*, and partially on chironomids. As *Hydropsyche* larvae are not typical for lentic waters, the specimen probably fed in the flowing section directly below the power plant or in the mouth of a tributary (HON-SIG – ERLENBURG & FRIEDL, 1999). There are various oral reports from fishermen of the An-nabrücke impoundment regarding juvenile sterlets, which probably did not originate from stocking measures. However, a natural reproduction has not been verified (HONSIG - ERLEN-BURG & FRIEDL, 1999). During a benthic sampling survey in the Völkermarkt impoundment in 1998 a skull of a 2.5 cm long specimen has been found (HONSIG - ERLENBURG & FRIEDL, 1999), which is the first and to this point the only proof of natural reproduction in the Austrian Drava. The larva was found 200 m downstream of a tributary in 8 m depth with a low flow velocity of 0.1 m/s over gravel bottom, overgrown with algae (HONSIG -ERLENBURG & FRIEDL, 1999). This habitat seems to be suitable for feeding also providing cover (protection from predation).

5.5.3.   Potential for population recovery

The various impoundments of the potamal Austrian Drava are rather short, with the flowing sections in between being even shorter. It seems unlikely that a self-sustaining population can be established. As the size of the caught fish rose continuously over the years after stocking, it can be assumed that all of these catches were from stocked fish and no natural reproduction took place. Three possible explanations might be given: (a) an insufficient number of spawners, (b) lack of spawning habitat and/or (c) poor adaptation of stocked fish to the natural environ-ment. However there might be two possible exceptions. As there are various reports of sight-

Figure 16: *Fragmentation of the Drava River, identified by different colours for each of the individual impoundments to visualize the migratory restrictions within the river caused by man-made barriers between which the species are trapped. Neighboring stretches with similar colours (f. e. blue) are connected. These fragmentations are super-imposed on Google Earth fotos. Numbers next to the coloured section indicate length of the impoundment in km. (Google/Geoimage Austria)*

ings of juvenile sterlets in the Annabrücke impoundment (HONSIG – ERLENBURG & FRIEDL, 1999) there could be, adequate spawning habitats. The finding of the sterlet larva in the Völkermarkt impoundment (HONSIG - ERLENBURG & FRIEDL, 1999) indicates that at least one successful spawning took place in the past. As most catches also occurred in these two impoundments, any restoration efforts in Carinthia should be concentrated on these water bodies. Nevertheless it has to be kept in mind that it is uncertain whether the sterlet ever had naturally been in existance in the Austrian Drava (Fig. 16).

5.6.   Mura River

5.6.1.   Historic records

MOJSISOVICS (1897) stated that the sterlet occurred in the Mura River up to Graz and reported a catch in the Andritzbach in 1890. It is likely that sterlets once were abundant in the stretch between Spielfeld and Bad Radkersburg. Statements that the species once occurred in the Austrian part of the Raab River are unlikely to be true due to the small size of the river in this area. However, it did occur in the Raab up to Körmend (WOSCHITZ, 2006). GLOWACKI (1885) described the Russian sturgeon from the mouth of the Mura River, while other reported the species from the area around Bad Radkersburg, but these remain doubtful (WOSCHITZ, 2006). ZAUNER et al. (2000) stated a possible occasional presence of the species in the Mura along the Austrian- Slovenian border. The catch of a ship sturgeon near Murakesztur on the Hungarian – Serbian border in 2005 (GUTI, 2006) proved the presence of this species in the Mura River. If the Russian sturgeon was present in the Mura River along the Austrian - Slovenian border in the past, it seems likely that also ship sturgeon, or even stellate or beluga sturgeons, were at least sporadically occuring.

5.6.2.   Stocking and recent catches

In 2001 the Mura River near Graz was stocked by the "Arbeiterfischereiverein Graz" ("Fish workers Association of Graz") with what was said to be sterlets. Unfortunately, no statistics, pictures or other confirmed data were available, to know the exact species stocked, the origin of the fish, or their size and quantity. In August 2005, members of the Institute of Hydrobiology and Aquatic Ecosystem Management found a dead Siberian sturgeon (Fig. A46) above the power plant Murau during a fish - ecological monitoring survey (WIESNER pers. comm.).   In 2010, two Siberian sturgeons, reported as sterlets (Fig. A47 & A48) were caught in the following nights near Spielfeld (Homepage of FV LEIBNITZ - www.fvl.at/cms/ 26.11.2010). Looking at the pictures of the fish and their length (78 & 79 cm total length) it might have been even the same specimen.

5.6.3.   Potential for population recovery

The Mura River downstream of Graz has many hydro - electric power dams but also other barriers which are not passable for sturgeons. The average length of the stretches between these barriers is approximately 5 km. Hydromorphological changes and channelization led to "rhithralization effects" (See Chapter Inn and Salzach Rivers) of the Mura River in the area

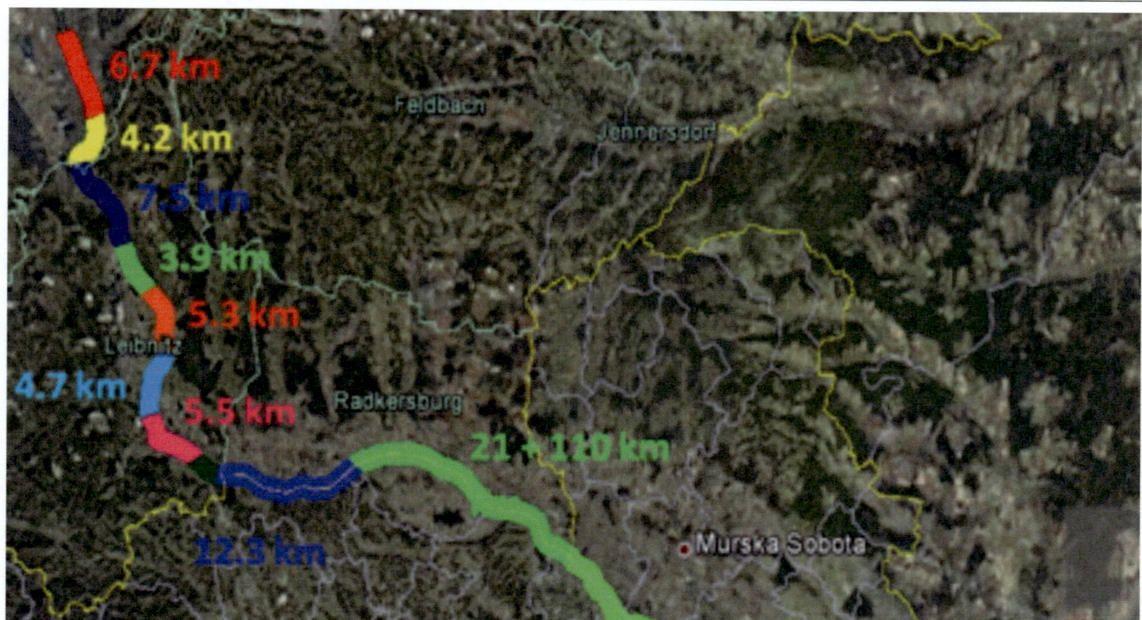

Figure 17: Fragmentation of the Mura River downstream of Graz and along the Slovenian - Austrian border, identified by different colours for each of the individual impoundments to visualize the migratory restrictions within the river caused by man-made barriers between which the species are trapped. Neighboring stretches with similar colours (f. e. blue) are connected. These fragmentations are super-imposed on Google Earth fotos. Numbers next to the coloured section indicate length of the impoundment in km. (Google/Geoimage Austria)

around Graz, rendering the habitat not suitable for sterlet but for many other highly endangered rheophilic species like the Danube salmon (*Hucho hucho*) or riffle dace (*Telestes souffia*). An interesting area is the 21 km long section downstream of the last migration barrier near the village Misselsdorf. This stretch along the Austrian – Slovenian border is still connected to around 110 km of the Mura River, 240 km of the Drava and 840 km of Danube River. Obviously, any efforts in this area on sturgeon rehabilitation must be a multilateral undertaking between countries sharing this watershed, including Austria, Croatia, Hungary, Slovenia and Serbia. The initial measures to be taken should concentrate on habitat improvements and reconstruction for the potamodromous sturgeon species. It can be assumed that many other species would also benefit from such habitat improvements, especially when coupled with fishing regulations and protection measures on key habitats. Thereafter, restocking of sterlet in the Mura River should be the next action. If this proves to be successful and if a ship sturgeon brood stock is available by then, this stretch of the Mura River might also serve as nursery habitat for restocked juvenile ship sturgeon. The problem of exotic sturgeon species might have certain relevance in this area. There are many hatcheries, aquaculture and recreational fishing lakes in lower Styria. Many of these keep or sell various sturgeon species like Siberian sturgeon, Russian sturgeon, white sturgeon, paddlefish or beluga sturgeon. It is more than likely that these fish are stocked (accidentally or on purpose) in the Mura River or its tributaries. For example one lake for recreational fishery, located directly next to the Sulm River, an important tributary of the Mura, has been densely stocked with Siberian sturgeons. The lake was flooded at least twice in the last years; so it seems very likely that a number of sturgeons escaped into the Sulm and migrated downstream into the Mura River (Fig. 17).

6. General overview on the status of sturgeons in Austria

6.1. Historic records

Despite intensive research it is not possible to present a complete picture of the exact historic distribution of the five sturgeon species known to occur in Austria. Therefore, two categories of maps have been prepared: Occurrence of the species verified by reported catches (**green**) and areas where the species was only stated by some authors or where occurrence seemed plausible regarding the distribution of other species with similar ecological demands (**yellow**).

The sterlet occurred throughout the Austrian Danube and in Bavaria upstream to Ulm. The species undertook spawning migrations up to 300 km in length (HOLCIK, 1989). Keeping the species` range and the migration pattern in mind and combining this information with records of catches, it can be assumed that it was common in the Austrian Danube, even if it might have been only temporarily present in some stretches. It also occurred in the Inn River along the Austrian - German border and in the Salzach River upstream to Laufen which can be seen as the limit of the sterlet distribution due to hydro-morphological changes in this area.

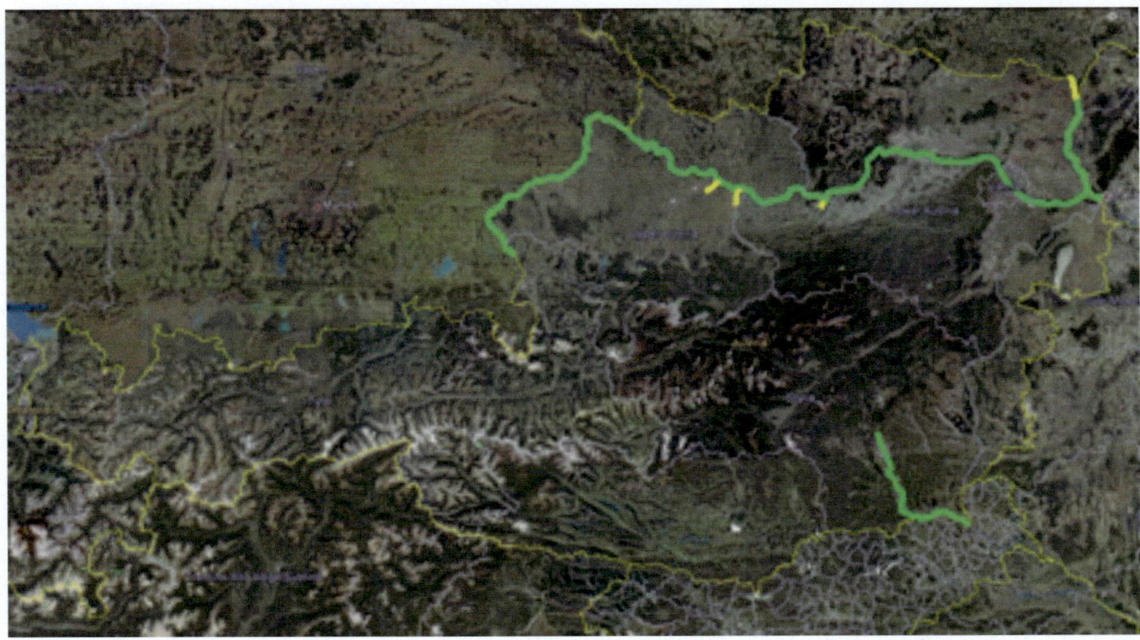

Figure 18: Historic distribution of the sterlet in Austrian rivers. Compiled from the information gathered from various sources (see material and methods). The distribution is super-imposed on Google Earth fotos. (Google/Geoimage Austria)

Although there are almost no reports from the rivers Traun, Enns and Ybbs, it can be assumed that at least within the lower 5 to 10 km were occasionally used by sterlets or other sturgeon species. The March River was inhabited by sterlets along the Austrian - Slovakian border and very likely also in the lower sections of the Thaya River. In the Mura a catch near Graz proves the occurrence of this species in this area, but due to hydro-morphological characteristics it can be expected that it was more abundant in the Austrian - Slovenian border section of the Mura. As there is no indication of the species presence in the Austrian Drava, the closest known sporadic occurrence being 70 km downstream near Maribor (WOSCHITZ, 2006), however, the sterlet probably never occurred in the Austrian Drava stretch (Fig. 18).

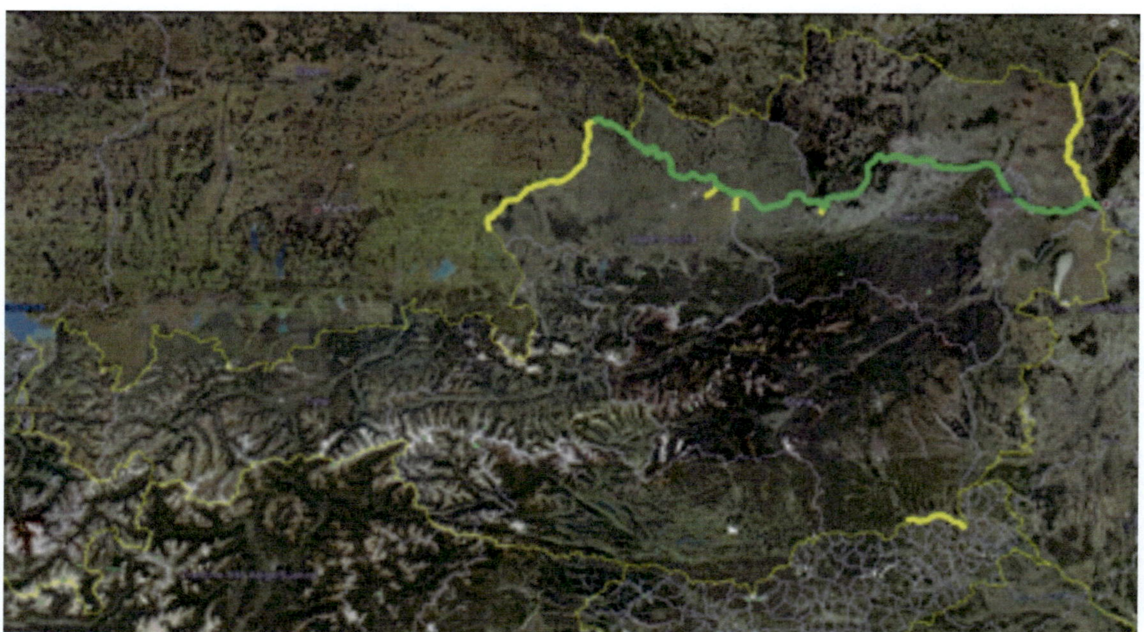

Figure 19: Historic distribution of the ship sturgeon in Austrian rivers. Compiled from the information gathered from various sources (see material and methods). The distribution is super-imposed on Google Earth fotos. (Google/ Geoimage Austria)

The ship sturgeon occurred in the entire stretch of the Danube River, as had been proven by catches near Vienna and Regensburg. Although not abundant in the 19th century (FITZINGER & HECKEL, 1839), it is likely that the species was already overfished by then but more numerous in earlier times. The situation is further complicated by fishermen making no distinction between juvenile ship sturgeons and sterlets and between adult ship sturgeons and Russian sturgeons. There are no reports of the species from other rivers in Austria but assessing the distribution of the large migratory species it seems possible that the ship sturgeon may have also occurred in the rivers Salzach, Inn, March and Mura. Also, the most downstream sections of the rivers Traun, Enns, Ybbs or Thaya might have had suitable habitats near their confluences with the Danube or the March, respectively (Fig. 19).

Russian sturgeons migrated up to the Bavarian Danube. It is also said to have been rare (FITZINGER & HECKEL, 1839), but this might be the same situation as described for the ship sturgeon. A catch near Schärding (BROD, 1980) shows the presence of this species in the Inn River; it probably also occurred in the Salzach River upstream to Tittmonig. In the March River it had been reported upstream to Hodonin. For the other tributaries, the same situation can be assumed as for the ship sturgeon. Various authors also assume the species to have occurred in the Mura River along the Austrian - Slovenian border (ZAUNER et al., 2000) (Fig. 20).

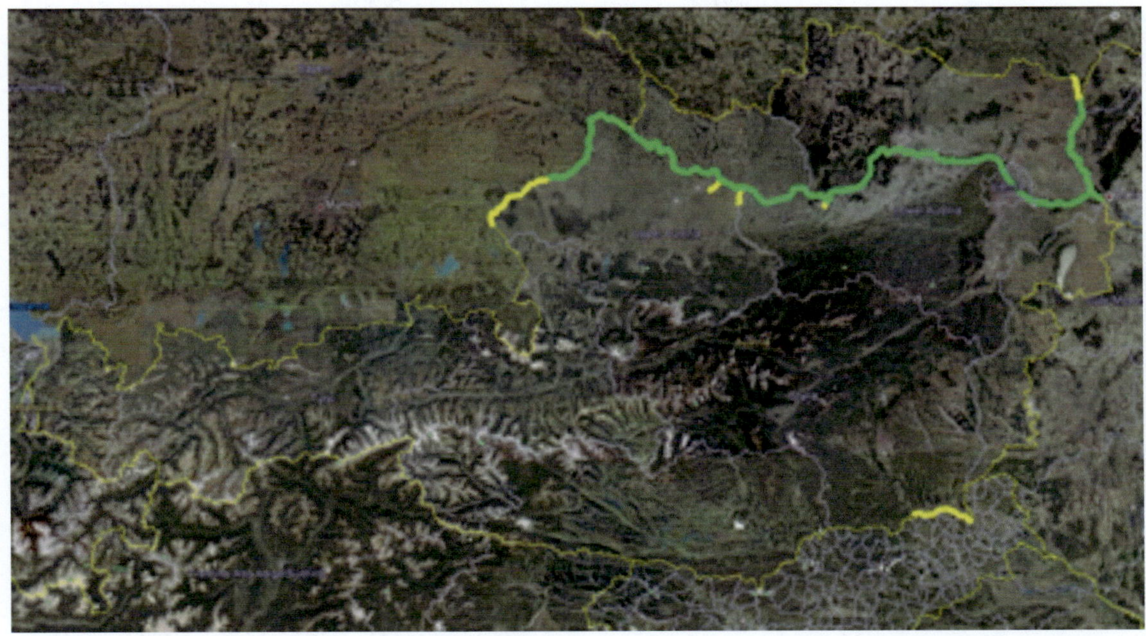

Figure 20: Historic distribution of the Russian sturgeon in Austrian rivers. Compiled from the information gathered from various sources (see material and methods). The distribution is super-imposed on Google Earth fotos. (Google/ Geoimage Austria)

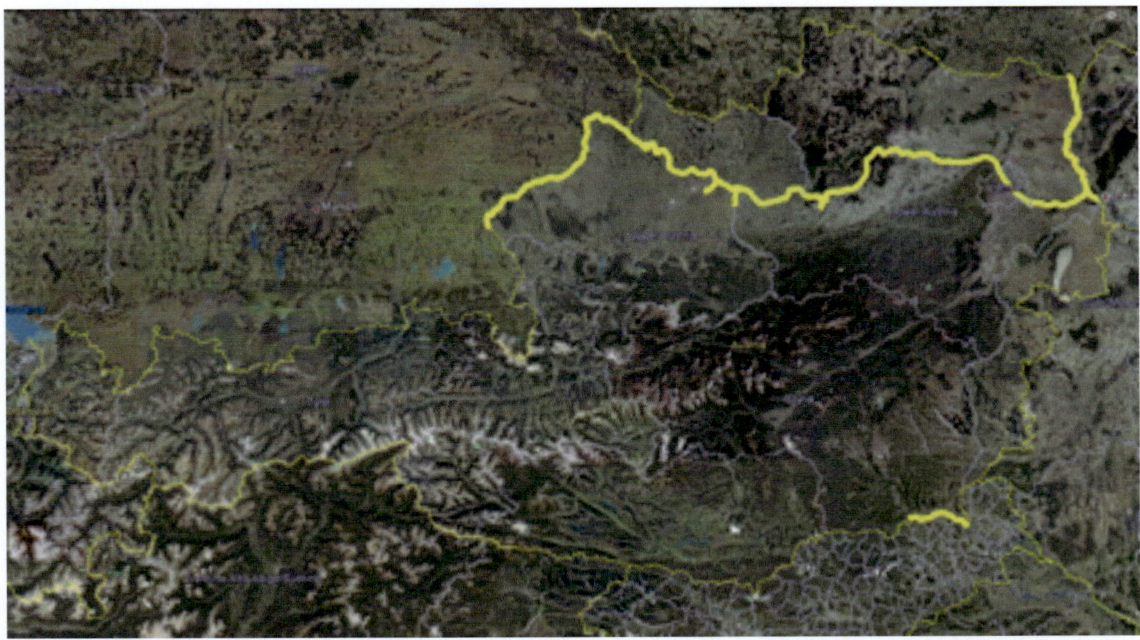

Figure 21: Historic distribution of the stellate sturgeon in Austrian rivers. Compiled from the information gathered from various sources (see material and methods). The distribution is super-imposed on Google Earth fotos. (Google/ Geoimage Austria)

There are no reports of stellate sturgeon catches in Austrian waters. However, there are two statements from the 19th century describing it rarely enters the Austrian Danube (FITZINGER & HECKEL, 1839) and rare occurrences in the Isar River in Bavaria (SIEBOLD, 1863). It probably was more abundant in historic times and used similar habitat and stretches as the other species. In the Ural River the spawning migration of the stellate sturgeon is considerably shorter than of Russian and beluga sturgeons (LAGUTOV & LAGUTOV, 2008), which might be similar in the Danube, being an indication that the species might have been less frequent than others (Fig. 21).

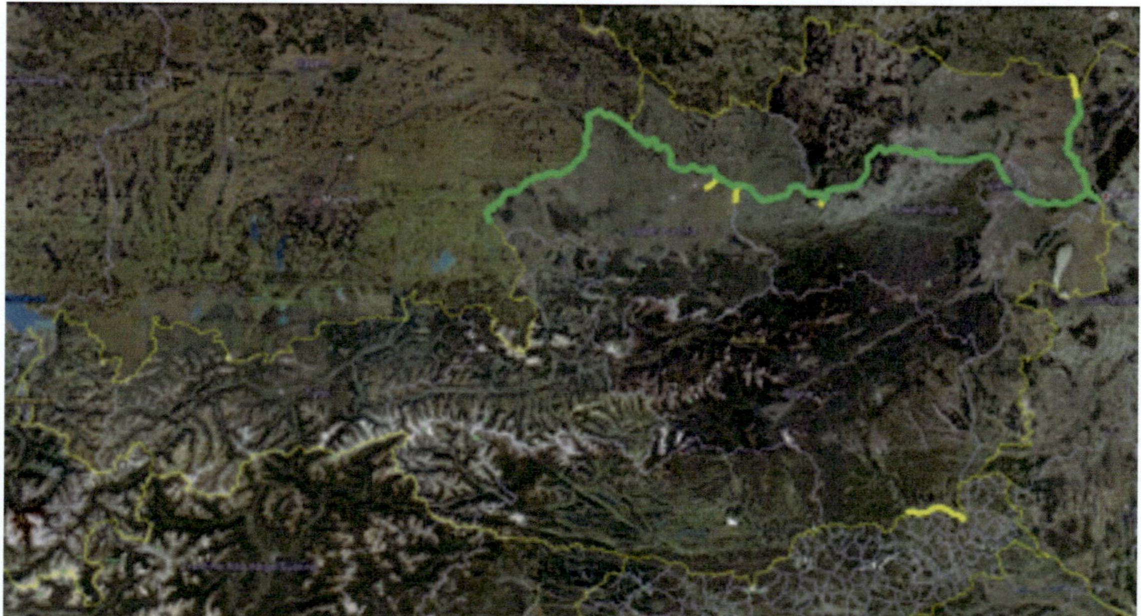

Figure 22: Historic distribution of the beluga sturgeon in Austrian rivers. Compiled from the information gathered from various sources (see material and methods). The distribution is super-imposed on Google Earth fotos. (Google/ Geoimage Austria)

The beluga sturgeon occurred throughout the Austrian Danube and the Inn River. In the Salzach River it migrated at least up to Tittmonig. Again, this species was stated as rare in Austrian waters in the 19th century (FITZINGER & HECKEL, 1839). In this case, the same authors noted that it was abundant in former times, which directly indicates overfishing in earlier centuries. For the March River, catches are reported as far upstream as Hodonin. For the other rivers, the situation might be similar to the Russian and ship sturgeon (Fig. 22).

6.2.   Stocking

Stocking was and still is conducted in various Austrian rivers, especially in the Danube and the Drava. Most stocked fish are sterlets, but there were also some stocking actions with extinct species like Russian sturgeon. The high number of illegally stocked or escaped alien sturgeon specimens is problematic due to possible hybridization or competition with native species. In the Drava a high number of sterlets was stocked between 1982 and 1995 in various impoundments. In the Danube stocking efforts with sterlets were concentrated near Vienna and in the Donau - Auen National Park area from 1994 to 2010.

6.3.   Recent catches

A total of 223 sturgeon specimens were recorded from Austrian rivers (Fig. 23) and were included in the database (PDB). It is, however, suspected that the real number, including unrecorded catches, is two to four times higher. About 90 reported catches were sterlets, half of them caught in the Danube and the other half in the Drava. Over 100 fish could not be identified due to the lack of pictures or other usable taxonomic information such as morphological characteristics. Probably most of these fish were sterlets and Siberian sturgeons. Twelve Siberian sturgeons have been reported. In river stretches not stocked with sterlets or not inhabited by a sterlet population, this species is by far the most abundant one in Austrian waters. Other species are caught to a lesser extent, for example white sturgeons, Russian sturgeons, various hybrids and paddlefish.

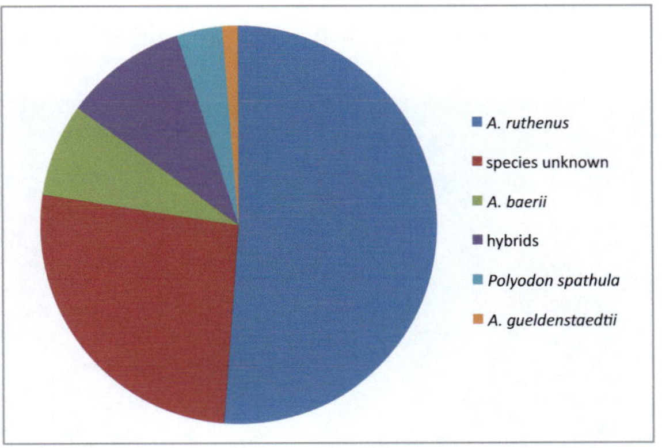

Figure 23: Sturgeon catches 1980 - 2011 in all Austrian rivers (n= 223)

In the Danube River, 83 specimens of various sturgeon species have been recorded (Fig. 24). About forty of these were sterlets, half of them being caught in the Aschach impoundment in the last 10 years. Catches of juveniles and adults are reported since the 1950s, indicating the last remaining self- sustaining population of this species in Austria. Other sterlet catches were from various river stretches: the most interesting catches originate from former populations in the 1980s and early 1990s near Linz, Wallsee, Altenwörth and Klosterneuburg. There is evidence of a natural repro-

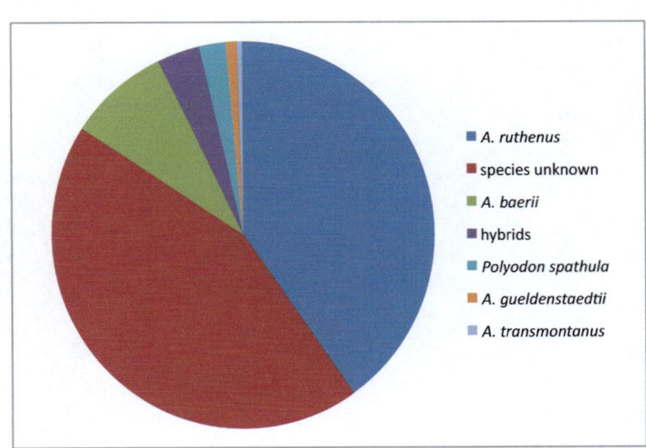

Figure 24: Sturgeon catches 1980 - 2011 in the Austrian Danube (n= 83)

duction near Klosterneuburg prior to the construction of the Freudenau power station. Intensive stocking in the early 21st century didn`t have any impact on catches upstream the Freudenau power station, whereas catches in the free flowing section downstream of Vienna increased after stocking. Many the caught specimens remained unidentified but it can be assumed that most of these were sterlets or Siberian sturgeons. Alien sturgeon species and hybrids have also been caught in nearly all Danube stretches, most fish being caught below the Altenwörth power station. An alarming discovery was the hybridization of native sterlets with alien Siberian sturgeon in the Aschach impoundment (LUDWIG et al., 2009), posing a potential threat to this last reproducing sterlet population (see Chapter Discussion - Alien species).

6.4.   Potential for population recovery

Although Austrian rivers suffer from various habitat alterations like channelization, migration barriers from hydro electric power stations or weirs, channel incisions, and other impacts., some stretches still have the potential to support a viable population of sterlets (see Chapter Discussion). Before any restocking programme is being considered, there is a need to assess the reasons for their disappearence.

Special attention has to be paid to the last remaining population below the Jochenstein

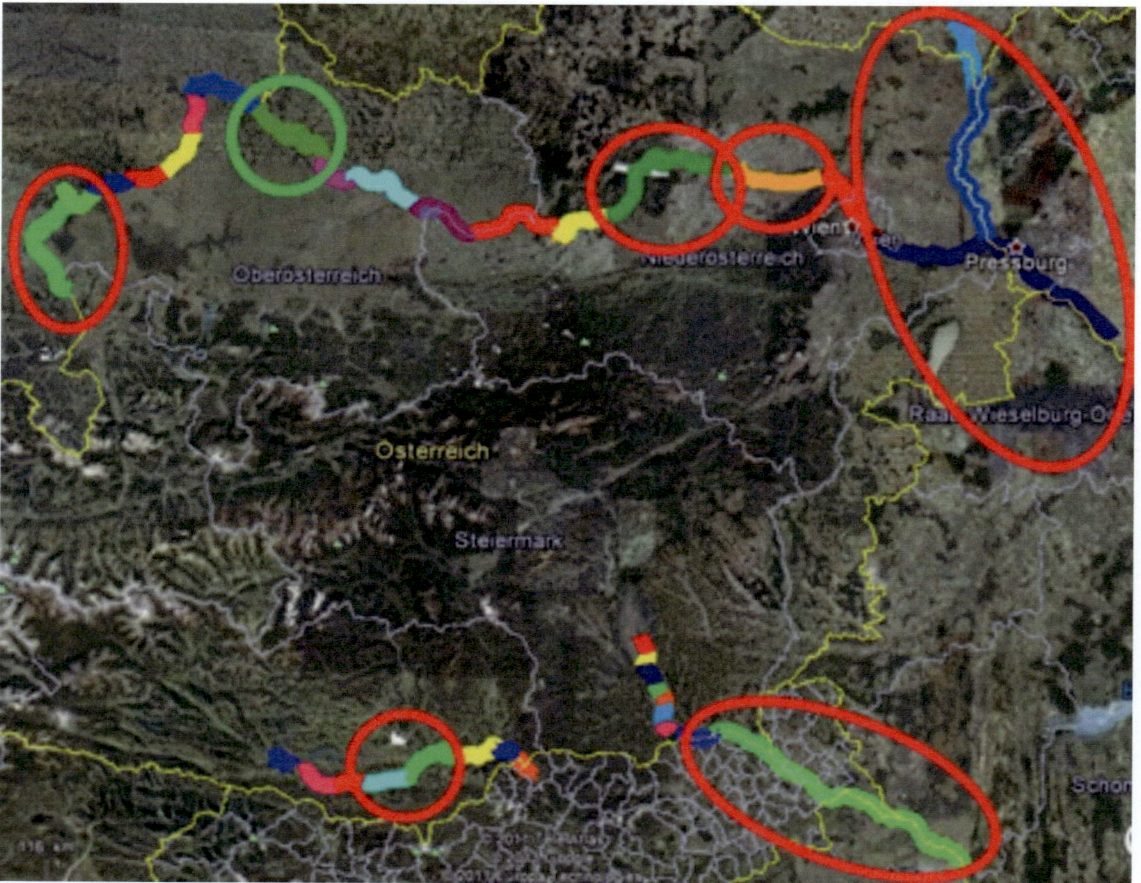

Fragmentation (highlighted by different colours, while similar colours show connected river stretches) and hotspots (red circles; Jochenstein area: green circle) for sturgeon restoration in Austria, super-imposed on Google Earth fotos. (Google/Geoimage Austria)

power station (see Fig. 25 stretch marked in green). Protection of this population from various influences like habitat loss and especially hybridization with alien sturgeon species are considered to be major risks. Mature brood fish from this stretch could be used to produce juveniles for restocking. Restocking activities should be accompanied by research on the habitat use of the Jochenstein population. Telemetric studies would be one possible way in order to identify and later reconstruct similar habitats in other Austrian river stretches.

The river system with the highest potential for re-introduction is the Danube downstream of Vienna, connected with the March and Thaya rivers, in total offering a free flowing river stretch of around 200 km in length. The latter two rivers show relatively little human impacts in terms of hydromorphological alterations and the Danube flows mostly through the protected area of the "Donau - Auen National Park". The system offers high habitat variability including both epi- and metapotamal characteristics. Successful restocking of sturgeons needs to be assessed, in terms of habitat availability, quality of progeny to be stocked and risk of losses by predators.

Other stretches that should be evaluated are the second free flowing Danube section in the famous Wachau valley in connection with the impoundment Altenwörth as well as the impoundment of Greifenstein. Sturgeon individuals caught in the latter exhibited very good condition, indicating that the available food resources seem to be adequate.

Another river with a high potential for re-colonisation, primarily for the sterlet and potentially for other species is the Mura River along the Austrian - Slovenian border. Even if there is only a very short stretch on Austrian territory, it is connected to over 1000 km of free flowing stretches of the rivers Mura, Drava and Danube. If adequate spawning and nursery habitats become available it could be restocked with fry and juveniles of sterlet and later even with juvenile ship sturgeons or anadromous sturgeon species once river connectivity via a fishpass to overcome the Iron Gate dams will become available. There is an ongoing discussion regarding a feasibility study for the construction of fish passes at these dams, future necessity if the countries are to comply with the EU Water Framework Directive. One remaining risk to re-established sturgeons will be the high density of sturgeon aquaculture plants and hatcheries in the area (which would have to be seriously controlled to avoid escapees) as well as the widespread distribution of alien sturgeon species within these watersheds.

In the Drava River, adequate sturgeon habitats are limited due to the numerous impoundments and only short stretches of free flow between them. The reports on juveniles in the impoundment of Annabrücke and the finding of a sterlet larva in the impoundment Völkermarkt (HONSIG - ERLENBURG & FRIEDL, 1999) are the only two indications of spawning after the stocking in the 1980s and 1990is. Therefore, any efforts in the Drava River on sterlet restoration should focus on these two river stretches.

The lower Salzach and Inn systems are also very small fragmented and show many alterations in their hydrology and morphology. The only interesting stretch is the connected environment of the Salzach and Inn of around 60 km length. All recently recorded sturgeon catches are from this area near the confluence of the Salzach River.

The most important stretches are along borders to other countries. Therefore all efforts regarding protection and re-introduction of any sturgeon species should be organized either bi- or multilateral and any actions taken should have pre- and post monitoring to prove the success, as stated in the Danube Sturgeon Action Plan (BLOESCH et al., 2006).

7. Discussion

At this time the sterlet is the most important sturgeon species in the Austrian Danube catchment as a reintroduction of other native sturgeon species is hampered by various reasons like lack of brood stock in captivity (ship sturgeon) or migration barriers further downstream (anadromous sturgeon species). Therefore the discussion will focus on the sterlet and especially on its habitat preferences as there is very little knowledge on habitat use in the Danube River.

7.1. Feeding habitat

According to catch data of this study, sterlets tend to stay at the upstream end of impounded sections with similar current patterns and conditions as in free - flowing rivers. Catches within the central parts of the impoundments are scarce. All pictures of caught specimens show that fish are in a very good condition. Therefore it can be assumed that at least intact feeding habitat and sufficient food supply are still available in the Austrian Danube. In the Danube sturgeons feed mostly in the main channel and occasionally in backwaters. Some authors even suggest the use of flooded lowlands (HOLCIK, 1989; BLOESCH et al., 2005; GUTI, 2008). As sturgeons are benthic fish, most of the diet consists of various benthic invertebrates. Some species like *A. sturio* show a high feeding selectivity in the wild while others like the closely related *A. oxyrinchus* rather shows opportunistic feeding habits (GESSNER, 2009). Therefore a straight forward interspecies transfer of behavioural characteristics is not suitable for all sturgeons (GESSNER, 2009). Most important prey organisms for the sterlet are *Chironomidae, Trichoptera, Ephemeroptera* and *Simuliidae* (HOLCIK, 1989). Additionally eggs of other fish are an important food source at least during certain times of the year, making out half of the stomach contents (HOLCIK, 1989). A recent study of sterlet diet in the Hungarian Danube showed that sterlets of all age classes prefer consuming chironomids, while bigger specimens also intensively feed on *Trichoptera* and gammarids of the genus *Corophium* (FIESZL et al., 2011). Plankton may also be important for the nutrition of fry of some sturgeon species. Therefore flooded lowlands might play an important role for healthy sterlet stocks. Mussels of the genus *Corbicula* have been found to be a temporally important food source for white sturgeon in North America (MILLER, 2004). These introduced mussels also occur in high densities in the Austrian Danube, especially in the impoundments. Additionally the various introduced Ponto - Caspian gammarid species might be a significant food source as the stocks increase throughout the Danube (MOOG, pers. comm.).

7.2. Wintering habitat

MOHR (1952) stated that sterlet and ship sturgeon spend the winter in the Danube in deep areas with clayey bottom without feeding while the beluga sturgeon prefers muddy bottom. Although these statements are questionable sterlets generally tend to stay in large shoals within deep sections of the river with good oxygen supply. A wintering site located directly in Budapest was discovered recently downstream a concrete bridge pier (GUTI, pers. comm.).

7.3.   Spawning habitat

Sturgeons spawn on gravel or rocky bottom in the main river channel. Some authors also suggest the use of floodplain sites flooded during high water levels (GUTI, 2006). The use of clayey and sandy substrates, as described by various earlier authors, has been ruled out for the Gulf sturgeon *Acipenser oxyrinchus desotoi* (SULAK & CLUGSTON, 1999). It can be assumed that the same may be true for other sturgeon species as the eggs would be at permanent risk of being covered by the fine substrate and limited oxygen supply. According to KIRSCHBAUM (2010) spawning and feeding habitats for sterlets are the same during low water levels, while spawning migrations tend to be longer during high water levels. HOCHLEITHNER (2004) stated that the spawning takes place in depths of 2 - 15 m over gravel beds and at flow velocities of 1.5 to 5 m/sec. Areas with quasi laminar flow show increased spawning activity (KHOROSHKO & VLASENKO, 1970), as the eggs are evenly distributed on the substrate, allowing good gas exchange at egg surfaces while also experiencing miminal sedimentation (ARNDT et al., 2006). Following spawning the spent adults migrate downstream to feed in bays and backwaters. HOLCIK (1989) stated that the spawning grounds of sterlets are located close to the wintering sites. Interpreting catch statistics of commercial fishermen in Hungary, it seems that optimal conditions for reproduction of sterlets and other litophilic spawners occur during moderate water levels in spring, as mortality of eggs and juveniles increases at very high water levels (GUTI, pers. comm.). Since sturgeon juveniles and fry are not very powerful swimmers (PEAKE, 2004) losses due to flood - induced drift might also play an important role. A traditional spawning site in the Szigetköz floodplain in Hungary was no longer usable for spawners after the construction of the Gabcikovo power plant dam (GUTI, 2008). The spawning sites were located in depths of around 8 m with gravel substrate (GUTI, pers. comm.). As the sterlets of the Jochenstein population are caught directly below the power plant most time of the year (ZAUNER, pers. comm.), they probably spawn on the rocky substrate in this area. According to KHOROSHKO & VLASENKO (1970), ide (*Leuciscus idus*) and asp (*Aspius aspius*) use the same spawning grounds as sturgeons do. These two species may be considered as indicators to search for potential sturgeon spawning habitats (GESSNER & BARTEL, 2000).

7.4.   Nursery habitat

Fry and juveniles remain at the spawning sites between rocks and stones whereas older juveniles are often encountered in sandy shallows (HOLCIK, 1989). The larva, found in a benthic sample in the Drava in 1998, was found 200 m downstream of a tributary in 8 m depth with a low flow velocity of 0.1 m/s over gravel bottom, overgrown with algae (HONSIG -ERLENBURG & FRIEDL, 1999). An experimental study researching the impact of various substrates on fry of Atlantic sturgeons showed significantly lower mortalities and better condition in fish reared over gravel substrate compared to fish reared over sand or without substrate (GESSNER et al., 2008).

7.5.   Predation

There is little data about predation on the different life stages of sturgeons. Generally fry and small juveniles are the life cycle stages most vulnerable to predation. Sub - adults and adults are much better protected against predators due to their size and their bony scutes. Young sterlets sometimes tend to feed from the surface, exposing their white belly. Stocking of a pond with 1000 fish of 5 cm total length resulted in a disaster, as most of the fish were eaten by seagulls (REICHLE, 1997). Other piscivorous birds like cormorants would likely have similar impacts, although observations under natural conditions have not yet been conducted. In North America predation of young lake sturgeons (*Acipenser fulvescens*) by crayfish (*Procambarus clarkii*) has been observed in a pond (AUER, 2004). Many Austrian rivers are inhabited by high densities of allochthonous crayfish (*Pacifastacus leniuscuslus* and *Orconectes limosus*). Invasive goby species have been observed near spawning sites of sturgeons in the Great Lakes and are suspected to feed on sturgeon eggs and larvae (AUER, 2004). The same goby species (bighead goby, *Neogobius kessleri*, round goby, *Neogobius melanostomus* and tubenose goby, *Proterohinus marmoratus*) actually are extremely abundant in the Austrian Danube and its tributaries and are assumed to have highly negative impacts on the native fish biocenosis. High densities of the European catfish (*Silurus glanis*) also pose a threat to juvenile sturgeon stocks both by direct predation and trophic competition. Negative impacts on Adriatic sturgeon stocks in various North Italian rivers are suspected to be caused by the increasing stock of catfish (PUZZI et al., 2009; BRONZI et al., 2011). The European catfish is a common species in the Austrian Danube and its stocks are likely to rise with increasing water temperatures due to climate change.

7.6.   Stocking and reintroduction

Stocking has been discussed with some controversy and major problems with the prevalent method of stocking sub- adult and adult fish have been described for various fish species and especially salmoniformes (for example BAARS et al., 2001; MARCHETTI & NEVITT, 2003; ARAKI et al., 2008; PINTER, 2008). Before any stocking actions with sturgeons are undertaken suitable habitats for all life stages should be identified. Stocking with sturgeons has been conducted in larger scales since the second half of the 20th century but failed in most cases due to various reasons. MARCHETTI & NEVITT 2003

The large stocking programs in the Caspian Sea, with millions of fingerlings released each year in the brackish estuaries increased the catches in the sea but did not significantly impact the catch numbers in the rivers. This indicates that the stocked fish do not have the homing fidelity as the imprinting on the natal river and their spawning sites did not take place (LAGUTOV & LAGUTOV, 2008).

Sturgeons reared in ponds over a longer period are trained to feed on pellet feeds. Own observations showed that older fish kept under such conditions often tend to ignore occasional supplied natural food items like fish, roe, crayfish, mussels or frozen chironomids. It can be assumed that many of these fish would poorly adapt to natural conditions. Recently, the need for adequate rearing of juvenile sturgeons for fitness for release has been assessed and published in a respective guidelines prepared by the respective global sturgeon experts working on the subject (see CHEBANOV et al., 2011).

The quantity to be released and the genetic variation within the cultured population also play an important role for a successful reintroduction in order to establish a self- sustaining population. Stocking of juveniles should be based on reproductions from a genetically clearly defined broodstock that encompass the complete genome of the natural population in order to avoid outbreeding and inbreeding depression. Further, the quantities to be stocked need to be carefully determined and must be based on solid studies on the productivity of the natural food web within the selected habitats. Only then one may be able to grossly estimate the carrying capacity of the entire river stretch. Subsequent monitoring studies involving a properly designed sampling programme must be initiated to allow to determine the natural mortality so that subsequent stocking programmes can be adjusted to the needs of a river-specific rehabilitation programme. Such considerations are a prerequisite to ensure the establishment of a founder population of sufficient size covering the natural genetic diversity (CHEBANOV et al., 2011). An isolated population should comprise of at least 50 and over a long term of at least 500 mature specimens (FRANKLIN, 1980). Later studies suggest even higher numbers of 1,000 to 5,000 individuals (JUNGWIRTH et al., 2003).

Stocking of sterlet fry in Hungary started in the 1980ies with several ten thousand specimens having been released each year. However, the impact on the numbers caught later remains questionable (GUTI, 2006; GUTI, 2008).

Table 6: Possible environmental factors and species characteristics potentially affecting the success of stocking various age classes (+ low/ ++ medium/ +++ high)

influencing factors	fry	juveniles	adults
predation	+++	++	
homing fidelity	+	++	+++
acclimation to natural environment	+	++	+++
food supply	+++	+	+

Three years after re-introduction of 2,000 juvenile sterlets with ~ 10 cm TL from Hungary into the Danube River in the impoundment Geisling in Germany, the first mature females were caught in the area (REICHLE, 1997). The fast growth and early maturity indicated that there was sufficient food supply to reach this stage. Nevertheless there are no indications of any successful natural reproduction in this area.

Stocking of juvenile and adult sterlets in various impoundments in the Austrian part of the Drava River in the 1980ies did have a significant impact on the catch statistics (HONSIG - ERLENBURG & FRIEDL, 1999), but as catch numbers decreased in recent years the species seems to vanish. Although the stocked fish partially have been fed with natural feeds and there are two hints of natural reproduction (HONSIG - ERLENBURG & FRIEDL, 1999), no self - sustaining population could be established. Possible reasons are the insufficient number of stocked fish or mature spawners, lack of habitat, missing homing fidelity or poor adaption to natural conditions.

Stocking in other Austrian river stretches increased catch numbers but also failed to result in any natural reproduction. As the caught fish were in very good condition they were apparently able to adapt to the natural food sources. Research on juvenile white sturgeons in the Kootenai River (USA), ranging from 23 to 72 cm in total length, for three years after release most of the juveniles seemed to thrive well in the given natural conditions, with an estimated survival rate of 60% in the first and 90% in the subsequent years (IRELAND et al., 2002; BURTSEV, 2009).

The introduction of sturgeon species outside of their natural range failed in most cases: for example the introduction of the sterlet in northern Germany (BREHM, 1910), the Russian sturgeon in the Baltic Sea and its tributaries (HOCHLEITHNER, 2004), the Siberian sturgeon in the Baltic Sea and the Volga (HOLCIK, 1989; PEGASOV, 2009). Although fish were caught on a regular basis (HOLCIK, 1989) they did not establish reproducing populations. A worldwide overview about various introductions, reintroductions, stocking and their success is given by WILLIOT et al. (2009).

With all these factors in mind it can be recommended to rear the fish for stocking in the same water as in the river and to feed them as much natural diet as possible. Whenever possible fry and fingerlings should be used for stocking and released near spawning sites to support "imprinting" on their home river and birthplace. Action plans for the Adriatic sturgeon in the Ticino River in Italy and the Guadalquivir in Spain suggest similar measures (ARLATI & POLIAKOVA, 2009; PUZZI et al., 2009; DOMEZAIN, 2009). Stocking should take place at least as long as the first stocked generation starts reproduction. When dealing with highly endangered species with only limited numbers of fish available for releasing (like the common sturgeon (*Acipenser sturio*) or the ship sturgeon), and also in river stretches with high risks of predation, it might be more promising to use bigger fish for stocking (KIRSCHBAUM et al., 2011). Ideally all stocking actions should be closely monitored, at least until the first generation of stocked fish starts spawning.

The reintroduction of the ship sturgeon in Austrian waters as proposed by various authors in and in various newspapers (ZAUNER, 1997; ANONYMOUS, 2004a; ANONYMOUS, 2004b; ANONYMOUS, 2004c; ANONYMOUS, 2004d; FRIEDRICH, 2009) is severely hampered by the lack of fish in captivity and the rarity of the species in the wild. The minimal population size and genetic bottleneck of this species in the Danube might have already been exceeded. One mature male ship sturgeon, caught near Mohács in 2010, was kept in the hatchery of a research institute in Hungary (GUTI, pers. comm.). Unfortunately this specimen died around the end of 2012 as it did not feed and adapt to life in captivity (GUTI, pers. comm.). Because of the bad condition it was not possible to perform alternative methods for artificial reproduction like dispermic androgenesis (GRUNINA, et al., 2009), or cryoconservation of the sperm (NORDHEIM, et al., 2001). As the species is nearly extinct in the Danube and potamodromous forms are also native in the Volga and Ural basin (LAGUTOV & LAGUTOV, 2008; KHODOREVSKAYA et al., 2009) it should be discussed to use these stocks for reintroduction if genetic differences between Danube and Caspian genotypes are minor. The IUCN guideline for reintroductions requires the used specimen to be as closely related to the native population as possible in order to increase the chances for optimal adaptation (IUCN/SSC RE - INTRODUTION SPECIALIST GROUP, 1995). Another possibility would be to put the genetic approach aside in favour of a species - conservation level approach. The Caspian and the Black Sea separated ~5 Mio years ago but reconnected for an indefinite time at the end of the last ice age around 10.000 years ago, giving sturgeon stock the possibility of genetic exchange between the two basins. Unfortunately the potamodromous Ural River stock of ship sturgeon is also close to extinction (LAGUTOV & LAGUTOV, 2008). The questions regarding upstream and downstream passage of sturgeons at the Iron Gates and Gabcikovo dams remain unresolved to this point. Therefore any efforts of reintroducing the anadromous sturgeon species in Austrian rivers are questionable. The still discussed possible potamodromous form of the Russian sturgeon seems to be even rarer than the ship sturgeon and might already be extinct.

7.7.   Alien sturgeon species in Austrian and other European waters

Alien sturgeon species and non- native stocks pose a potential threat to autochthonous sturgeon stocks through possible hybridization, competition or disease transfer. The encounter of mature Siberian sturgeons and hybrids between Siberian sturgeons and sterlets in the impoundment of Aschach (LUDWIG et al., 2009) verified the seriousness of this issue and disproved other authors questioning the Siberian sturgeons ability to adapt to conditions in middle European rivers (REICHLE, 1997). Some authors suggested the caught hybrids to be accidentally stocked fish (GESSNER, 2009). Although this so - called sibster hybrid can be found in hatcheries, it is very rare and unlikely to be released in such high numbers. A nearly ripe female of the Siberian sturgeon was also caught in the Garonne River in France (WILLIOT et al., 2009), potentially posing a threat to the last population of *A. sturio*. Although it is unlikely that the Siberian sturgeon will be able to establish self-sustaining populations in European waters, this potential cannot be fully excluded at regional level considering the rising number of stocks of mature fish anticipated to occur during the next years. Introduction of stellate sturgeons in the Aral Sea contributed to the collapse of native ship sturgeon stocks due to various introduced parasites (HOCHLEITHNER, 2004). Various viral diseases found in hatcheries raising white sturgeon, shovelnose sturgeon and Russian sturgeon can also pose a serious threat to wild populations (HOCHLEITHNER, 2004; VAN EENEENNAAM et al., 2004). Numerous paddlefish have been caught in several countries along the Danube, especially in the lower parts and the possibility

Table 7: Assumed trading level for different sturgeon species (juveniles and adults) in Austria from 1990 to 2012 (+ rare/ ++ regularly/ +++ frequent)

species	juvenile	adult
A. ruthenus	+++	+
A. gueldenstaedtii	+++	++
A. stellatus	++	+
H. huso	+	+
A. baerii	+++	+++
A. naccarii	+	+
A. transmontanus	+	++
A. oxyrinchus	+	
P. spathula	+	
hybrids	++	++

of its naturalization was discussed by some authors (HOLCIK, 2006; SIMONOVIC et al., 2006). In Germany high numbers of various allochthonous sturgeon species have been caught in the last years (ARNDT et al., 2000; WIESNER et al., 2010) posing a threat to conservation programs for the common European sturgeon (*Acipenser sturio*) in the North Sea and the Atlantic sturgeon (*Acipenser oxyrinchus*) in the Baltic Sea. Most caught specimens are Siberian sturgeons, Russian sturgeons or of unknown identity (ARNDT et al. 2001). Like in Austria it can be assumed that the number of unrecorded catches is several times higher. As sturgeon production levels increase worldwide, it is clear that the number of alien sturgeons within various river systems will rise if no precautionary actions are taken. Although this development cannot be stopped or reversed in the near future, various measures to deal with this topic have been discussed (ARNDT et al., 2001, FRIEDRICH, 2009). Sensitization of fishermen, hatchery and garden pond owners etc. regarding the risks of releasing allochthonous sturgeons into wild water bodies can be seen as a first step. At this point caught fish are often only distinguished by fishermen between "sterlet" (long snout) and "sturgeon" (short snout). Other possible future measures include stricter trade control, higher fines for illegal stocking and encouraged "taking" of exotic sturgeons by fishermen once they are able to differentiate them from sterlets (FRIEDRICH, 2009).

7.8.   Management issues

Due to its conservation status it is recommended to protect the sterlet in all Austrian provinces. To this point the minimum size limits are even too small (Tab. 8) to guarantee at least one reproduction cycle (FRIEDRICH, 2009). This subject is under discussion in Upper Austria (PILGERSTROFER, pers. comm.) and a total ban of sterlet fishing will be implemented in 2014 (BERG, pers. comm.). Fortunately, most recreational fishermen tend to release caught specimens alive. A commercial fishery on the Austrian Danube is nearly non - existing, as there are only handful fishermen using nets to catch various cyprinids and other fish. The by-catch sometimes comprises sterlets, which are released in most cases.

Table 8: Fishing regulations for sterlet in Austria, listed for different province with indications on minimum size limits

Province	minimum size limit (cm TL)	closed season	source
Upper Austria	45	1. May - 30. June	OÖ. LANDESFISCHEREIVERBAND, 2004
Lower Austria	45	1. May - 30. June	NÖ. LANDESREGIERUNG, 2001
Vienna	protected	protected	WIENER LANDESREGIERUNG, 2008
Styria	50	1. April - 30 June	STEIERMÄRKISCHE LANDESREGIERUNG, 2000
Carinthia	40	1. January - 30. June	KÄRNTNER LANDESREGIERUNG, 2001
Salzburg	protected	protected	SALZBURGER LANDESREGIERUNG, 2003

7.9.   Conclusions

There is little knowledge of the habitat use by sturgeons in the Danube and other Austrian river systems. Future conservation actions require sufficient information on the characteristics of key habitats. Therefore, the last remaining population in Jochenstein and its habitat should be studied intensively in order to be able to identify the habitats and to search for similar areas in other river stretches. Appropriate methods to determine habitat use would include implants of radiotelemetric or hydroacoustic transmitters in a number of fish to monitor their migration patterns and exact locations within the Danube. At this point there are no reports on catches downstream of Obermühl, indicating the fish may only use a 25 km long section of the upper impoundment. The recent catch of two marked specimens (ZAUNER, pers. comm.) stocked in the river Schwarze Laber in Germany further showed the attractiveness of this stretch for sterlets, as these specimens migrated over 150 km downstream, passed several power plants on their way and finally stayed in the upstream section of the Aschach impoundment. Hence, the Jochenstein population can be seen as the key to sturgeon conservation and restoration in Austria.

The factors which caused extinction of the sterlet in many stretches remain still unclear. One key factor seems to be the loss of spawning places, like upstream of Vienna. Here, spawning places were indicated by catches of juveniles, but finally lost after the construction of the Freudenau power plant. This key factor coupled with already low numbers of mature adults may have led to the vanishing of the sterlet in most river sections.

The construction of artificial spawning grounds proved to be successful below the power plants Krasnodar and Fedorovskiy in the Kuban River. They were built 300 to 1500 meters below the dams but unfortunately after three years they diminished due to siltation and deposition (CHEBANOV et al., 2008). Positive results downstream of the Volgograd dam in the Volga River also showed the effectiveness of such sites (MALTSEV, 2009). The layer should be at least 20 cm thick and composed of gravel with a diameter of 3 - 10 cm (KOTENEV, 2009). Construction of artificial spawning sites should be considered as an option in the Austrian Danube, especially below the Freudenau power plant. Regarding sustainability such sites should be located in areas with natural gravel sediments or constructed with gravel in a grain diameter big enough in order to avoid deposition of the gravel banks. Using hydrodynamic models and sediment regime monitoring, the gravel banks can be observed and constructed in a way to maximize their efficiency and durability.

8. References

Aigner, J.; Zetter, J.T.M., 1859: Salzburgs Fische. (Fishes of Salzburg). Jahresbericht des vaterländischen Museums Carolino- Augusteum der Landeshauptstadt Salzburg, pp. 72 – 92. [In German].

Anonymous, 1884: Rechenschaftsbericht des oberösterreichischen Fischerei- Vereines über das Jahr 1883. (Activity report of the Fisheries association of upper Austria for the year 1883). Mitteilungen des oberösterreichischen Fischerei- Vereines Jg. 4, Nr. 14, pp. 87 - 91. [In German].

Anonymous, 2000: Sterlet im Traunsee. (Sterlet in the Traun Lake). Fisch und Fang, Nr. 12/2000, pp. 14. [In German].

Anonymous, 2001: Der Stör *Acipenser sturio* L. Fisch des Jahres 2001. (The sturgeon: fish of the year 2001). Verband deutscher Sportfischer e. V. (editor), Offenbach, Germany, 86 pp. [In German].

Anonymous, 2004a: Stör wieder ansiedeln - Kaviar aus der Donau. (Restocking of sturgeons – caviar from the Danube). Kronenzeitung Oberösterreich, Linz, Austria, 6.02.2004. [In German].

Anonymous, 2004b: FH Salzburg mit Stör - Studie. Der Standard, Wien, Austria, 31.01.2004. [In German].

Anonymous, 2004c: Stör soll in die Donau zurückkehren. (Sturgeons to return into the Danube). Salzburger Nachrichten, Salzburg, Austria, 6.02.2004. [In German].

Anonymous, 2004d: Nach 70 Jahren Störe in der Donau. (After 70 years sturgeons return to the Danube). Salzburger Wirtschaft, Salzburg, Austria, 30.01.2004. [In German].

Anonymous, 2005: Rotauge & Blaunase. O.Ö. (= upper Austrian) Nachrichten, Linz, Austria, 10.12.2005. [In German].

Antipa, G., 1905: Die Störe und ihre Wanderungen in den europäischen Gewässern mit besonderer Berücksichtigung der Störe der Donau und des Schwarzen Meeres. (Sturgeons and their migrations in the European waters with special attention of the Danube and Black Sea sturgeons.). Bericht an den internationalen Fischerei - Kongreß in Wien 1905. [In German].

Araki, H.; Berekijian, B. A.; Ford, M. J.; Blouin, M. S., 2008: Fitness of hatchery-reared salmonids in the wild. – Evolutionary Applications, pp. 342 - 356.

Arlati, G.; Poliakova, L., 2009: Chapter 14 - Restoration of Adriatic Sturgeon *(Acipenser naccarii)* in Italy: Situation and Perspectives. pp. 237 - 245. In: Carmona, R.; Domezain, A.; Garcia - Gallego, M.; Hernando, J. A.; Rodriguez F.; Ruiz – Rejon, M., (Eds), 2009: Biology, Conservation and Sustainable Development of Sturgeons. Heidelberg, Dordrecht, London, New York: Springer Verlag. VIII, 668 pp. (ISBN: 987-3-642-20610-8)

Arndt, G. M.; Gessner, J.; Anders, E.; Spratte, S.; Filipak, J.; Debus, L.; Skora, K., 2000: Predominance of exotic and introduced species among sturgeons captured from the Baltic and North Seas and their watersheds, 1981 - 1999. Boletin Instituto Espanol de Oceanografia 16, pp. 29 -36.

Arndt, G. M.; Gessner, J.; Spratte, E., 2001: Doch noch Störe in Deutschland? - Fänge von einheimischen und nicht einheimischen Stören in mitteleuropäischen Küstengewässern. (Are there still sturgeons in Germany? - catch of endemic and exotic sturgeons in central European waters). pp. 50 – 62. In: Anonymous, 2001: Der Stör *Acipenser sturio* L. Fisch des Jahres 2001. (The sturgeon: fish of the year 2001). Verband deutscher Sportfischer e. V. (editor), Offenbach, Germany, 86 pp. [In German].

Arndt, G. M.; Gessner, J.; Bartel, R., 2006: Characteristics and availability of spawning habitat for Baltic sturgeon in the Odra River and its tributaries. pp. 89 – 99. In: Gessner, J., 2009: Prerequisites for the remediation of the indigenous sturgeons *Acipenser sturio* Linnaeus 1758 and *A. oxyrinchus* Mitchill 1815 in river systems of northern Germany - development of methods and contributions towards scientific criteria and concepts for a recovery management plan. Humboldt University, Berlin, Germany, 216 pp. (ISBN: 978-3-86624-448-1)

Auer, N., 2004: Chapter 12 - Conservation. pp. 252 - 276. In: LeBreton, G.T.O.; Beamisch, F.W.H.; McKinley, R.S., (Eds.). 2004. Sturgeons and Paddlefish of North America. Dordrecht/ Boston/London: Kluwer Academic Publishers, XII, 323 pp. (ISBN 1-4020-2832-6). (Fish and Fisheries Series, vol. 27).

Baars, M.; Mathes, E.; Stein, H.; Steinhörster, U., 2001: Die Äsche *Thymallus thymallus*. (The grayling *Thymallus thymallus*). Die Neue Brehm-Bücherei, Westarp Wissenschaften, Hohenwarsleben, Germany, 128 pp. (ISBN-13-978-3894328818) [In German].

Bacalbaca –Dobrivici, N.; Holcik, J., 2000: Distribution of *Acipenser sturio* L., 1758 in the Black Sea and its watershed. Boletin Instituto Espanol de Oceanografia 16, pp. 37 - 41.

Balon, E. K., 1968: Einfluß des Fischfangs auf die Fischgemeinschaften der Donau. (Influence of fish catching on the fish communities in the Danube River). Arch. Hydrobiol./Suppl. XXXIV (Donauforschung III), Stuttgart, pp. 228 - 249. [In German],

Bartosiewicz, L.; Bonsall, B., 2008: Complementary taphonomies: Medevial sturgeons from Hungary. Archéologie du Poisson, 30 ans d`Archéo - Ichtyologie au CNRS, Antibes, pp. 35 - 45.

Bloesch, J.; Jones, T.; Reinartz, R.; Striebel, B., (Eds.), 2005: Action Plan for the Conservation of the Sturgeons (Acipenseridae) in the Danube River Basin. Convention on the Conservation of European Wildlife and Natural Habitats, Strasbourg, France, 121pp. (ISBN 978-92-871-5992-2)

Brehm, A. G., 1910: Das Leben der Tiere - Die Kriechtiere - Die Fische. (The life of animals – reptiles and fishes). Deutsche Buch Gemeinschaft, Berlin, Page 400. [In German].

Brod, W. M., 1980: Historische Streiflichter auf Fisch und Fischerei. (Historic flashlights on fish and fishery). Im Dienste der Bayrischen Fischerei- 125 Jahre Landesfischereiverband Bayern e. V.: pp. 173 – 237. [In German].

Bronzi, P.; Castaldelli, G.; Cataudella, S.; Rossi, R., 2011: Chapter 16 - The Historical and Contemporary Status of the European Sturgeon, *Acipenser sturio* L., in Italy. pp. 227 – 241. In: Williot, P.; Rochard, E.; Desse - Berset, N.; Kirschbaum, F.; Gessner, J., (Eds.), 2011: Biology and Conservation of the European Sturgeon Acipenser sturio L. 1758 - The Reunion of the European and Atlantic sturgeons. Heidelberg, Dordrecht, London, New York: Springer Verlag. VIII, 668 pp. (ISBN: 987-3-642-20610-8)

Burtsev, I. A., 2009: Chapter 22 - Towards the Definition of Optimal Size - Weight Standards of Hatchery - Reared Sturgeon Fry for Restoration. pp 359 – 368. In: Carmona, R.; Domezain, A.; Garcia - Gallego, M.; Hernando, J. A.; Rodriguez F.; Ruiz – Rejon, M., (Eds),

2009: Biology, Conservation and Sustainable Development of Sturgeons. Heidelberg, Dordrecht, London, New York: Springer Verlag. VIII, 668 pp. (ISBN: 987-3-642-20610-8)

Busnita, T., 1967: Die Fischerei und Fischereiwirtschaft. (Fishery and fish industry). In: Liepolt, R. (Editor), 1967: Limnologie der Donau. (Limnology of the Danube River). E. Schweizerbart'sche Verlagsbuchhandlung, Stuttgart, Germany, 648 pp. (ISBN 978-3510990528) [In German]

Carmona, R.; Domezain, A.; Garcia - Gallego, M.; Hernando, J. A.; Rodriguez F.; Ruiz – Rejon, M., (Eds), 2009: Biology, Conservation and Sustainable Development of Sturgeons. Heidelberg, Dordrecht, London, New York: Springer Verlag. VIII, 668 pp. (ISBN: 987-3-642-20610-8)

Chebanov, M. S.; Galich, E. V.; Ananyev, D.V., 2008: Strategy for conservation of sturgeon under the conditions of the Kuban River flow regulation. pp. 70-82. In: Rosenthal, H., Bronzi, P., Poggioli, C. & M. Spezia, (Eds.) 2008: Passages for fish - overcoming barriers for large migratory fish. World Sturgeon Conservation Society Special Publication No. 2, Books on Demand GmbH, Norderstedt, –Germany, IV, 165 pp. (ISBN 978-3-8370-6142-0)

CITES (Convention on international Trade in endangered Species of wild Flora and Fauna, 2012: Appendices I, II and III valid from 3 April 2012, pp. 27. http://www.cites.org/eng/app/appendices.php

Domezain, A., 2009: Chapter 26 - Main Steps and Proposals for a recovery Plan of Sturgeon in the Guadalquivir River (Spain). pp 423 – 452. In: Carmona, R.; Domezain, A.; Garcia - Gallego, M.; Hernando, J. A.; Rodriguez F.; Ruiz – Rejon, M., (Eds), 2009: Biology, Conservation and Sustainable Development of Sturgeons. Heidelberg, Dordrecht, London, New York: Springer Verlag. VIII, 668 pp. (ISBN: 987-3-642-20610-8)

Eberstaller, J.; Pinka, P.; Honsowitz, H., 2001: Überprüfung der Funktionsfähigkeit der Fischaufstiegshilfe am Kraftwerk Freudenau. (Checking the functionality of the fishpass at the powerstation Freudenau). Schriftenreihe der Forschung im Verbund Band 72, Vienna, Austria, 95 pp. [In German].

Eder, K., 2003: Sterletprojekt bis 2005 verlängert. (Sterlet project expanded until 2005). Österreichs Fischerei, 56, pp. 170. [In German].

European Council, 1992: Council Directive 92/43/EEC of 21 May 1992 on the conservation of natural habitats and of wild fauna and flora.

Fieszl, J.; Bogacka-Kapusta, J.; Kapusta, A.; Szymanska, U.; Martyniak, A., 2011: Feeding ecology of sterlet *Acipenser ruthenus* L. in the Hungarian section of the Danube. Arch. Pol. Fish. 19, pp. 105 - 111.

Fischer, H., 1952: Die Störe. (The sturgeons). Österreichs Fischerei 5/1952, pp. 251. [In German].

Fitzinger, J. A.; Heckel, J., 1836: Monographische Darstellung der Gattung *Acipenser*. (Monograph on the genus *Acipenser*). Annalen des Wiener Museums der Naturgeschichte, 1. Band, Vienna. Austria, 65 pp. [In German].

Franklin, I. R., 1980: Evolutionary change in small populations. pp. 135 – 149. In Wilcox, E. A. (Editor), 1980: Conservation Biology - An Evolutionary Ecological Perspective. Sinauer Associates Inc, Sunderland, United States, 395 pp. (ISBN 978-0878938001)

Frauenfeld, G. R. v., 1871: Die Wirbeltierfauna Niederösterreichs. (The vertebrate fauna of lower Austria). Blätter des Vereins für Landeskunde von Niederösterreich 5, pp. 108- 123. [In German].

Freudlsperger, H., 1936: Kurze Fischereigeschichte des Erzstiftes Salzburg. (Short history of fisheries in the "Erzstift Salzburg"). Mitteilungen der Gesellschaft für Salzburger Landeskunde. Jg. 76, 77, part 1 + 2. [In German].

Friedrich, T., 2009: Störartige in Österreich- Chancen und Perspektiven im Freigewässer. (Sturgeon-like fishes in Austria – chances and perspectives in open waters). Österreichs Fischerei, 62, pp.250 - 258. [In German].

Gamlitschek, A., 1897: Die Stadt Tulln und ihre Fischwässer von ehegestern und heute. (The city of Tulln and its fisheries in the past and present). In: Mitteilungen des österreichischen Fischerei- Vereins 17, pp. 26 - 33. [In German].

Gessner, J.; Bartel, R., 2000: Sturgeons spawning grounds in the Odra River tributaries - a first assessment. pp. 77 – 88. In: Gessner, J., 2009: Prerequisites for the remediation of the indigenous sturgeons *Acipenser sturio* Linnaeus 1758 and *A. oxyrinchus* Mitchill 1815 in river systems of northern Germany - development of methods and contributions towards scientific criteria and concepts for a recovery management plan. Humboldt University, Berlin, Germany, 216 pp. (ISBN: 978-3-86624-448-1)

Gessner, J.; Kamerichs, C. M.; Kloas W.; Wuertz, S., 2008: Behavioural and physiological responses in early life phases of Atlantic sturgeon (*Acipenser oxyrinchus* Mitchill 1815) towards different substrates. pp. 61 – 76. In: Gessner, J., 2009: Prerequisites for the remediation of the indigenous sturgeons *Acipenser sturio* Linnaeus 1758 and *A. oxyrinchus* Mitchill 1815 in river systems of northern Germany - development of methods and contributions towards scientific criteria and concepts for a recovery management plan. Humboldt University, Berlin, Germany, 216 pp. (ISBN: 978-3-86624-448-1)

Gessner, J., 2009: Prerequisites for the remediation of the indigenous sturgeons *Acipenser sturio* Linnaeus 1758 and *A. oxyrinchus* Mitchill 1815 in river systems of northern Germany - development of methods and contributions towards scientific criteria and concepts for a recovery management plan. Humboldt University, Berlin, Germany, 216 pp. (ISBN: 978-3-86624-448-1)

Glowacki, J. (Editor), 1885: Die Fische der Drau und ihres Gebietes. 16. Jahresbericht des Steierm. Landes Untergymnasiums zu Pettau,. Commissions-Verlag von A. Pichler's Witwe & Sohn, Pettau, Austria, 18 pp. [In German].

Grunina, A. S.; Recoubratsky, A. V.; Barmintsev, V. A.; Vasi`eva E. D.; Chebanov, M.S., 2009: Chapter 11 - Dispermic Androgenesis as a Method for Recovery of Endangered Sturgeon Species. pp. 187 – 204. In: Carmona, R.; Domezain, A.; Garcia - Gallego, M.; Hernando, J. A.; Rodriguez F.; Ruiz – Rejon, M., (Eds), 2009: Biology, Conservation and Sustainable Development of Sturgeons. Heidelberg, Dordrecht, London, New York: Springer Verlag. VIII, 668 pp. (ISBN: 987-3-642-20610-8)

Guti, G., 2006: Past and present status of sturgeons in Hungary. IAD (International Association for Danube Research) Conference 2006 Szentendre, Hungary, 21- 24. August 2012.

Guti, G., 2008: Past and present status of sturgeons in Hungary and problems involving their conservation. Fundam. Appl. Limnol./Arch. Hydrobiol., Suppl. 162., Large Rivers Vol. 18. No.1-2, Pages 61 - 79.

Haidvogl, G.; Waidbacher, H., 1997: Ehemalige Fischfauna an ausgewählten österreichischen Fließgewässern. (Historic fishfauna of selected Austrian rivers). Department of Water, Atmosphere and Environment - Institute of Hydrobiology and Aquatic Ecosystem Management, University of Natural Resources and Applied Life Sciences, Vienna, Austria, 86 pp. [In German].

Heckel, J., 1854: Die Fische der Salzach. (The fishes of the Salzach). Verh. zool.-bot. Ver. Wien Austria, 4, Pages 189 - 196. [In German].

Heckel, J. & R. Kner, 1857: Die Süßwasserfische der österreichischen Monarchie mit Rücksicht auf die angränzenden Länder. (The freshwater fish of the Austrian monarchy with consideration of the neighbouring countries). Verlag von Willhelm Engelmann, Leipzig. 388 pp. [In German].

Heinrich, A., 1856: Mährens und k.k. Schlesiens Fische, Reptilien und Vögel. (Fish, reptiles and birds of Mähren and Schlesien). Brünn, Czech Republic, 2 pp. [In German].

Hensel, K. Holcik, J., 1997: Past and current status of sturgeons in the upper and middle Danube River. – Env. Biol. Fish. 48, pp. 185 - 200.

Hirmer, M., 2011: Sterlet soll wieder heimisch werden. (An intended come-back for the sterlet). Fischer & Teichwirt , 62 (11), pp. 413. [In German].

Hochleithner, M., 2004: Störe- Biologie und Aquakultur. (Sturgeons- biology and aquaculture). AquaTech Publications, Kitzbühel, Austria, 228 pp. (ISBN 3-9500968-2-5) [In German].

Holcik, J.; Bastl, I.; Ertl, M.; Vranowsky, M., 1981: Hydrobiology and ichthyology of the Czechoslovak Danube in relation to predicted changes after the construction of the Gabcikovo-Nagymaros River Barrage System. – Práce lab. rybar. hydrobiol 3, pp. 19 - 158.

Holcik, J. (Editor), 1989: The Freshwater Fishes of Europe - Vol. 1, Part II General Introduction to Fishes/Acipenseriformes. AULA- Verlag GmbH, Wiesbaden, Germany 469 pp. (ISBN 3-89104-431-3)

Holcik, J., 1995: Acipenseriformes. pp. 372 - 397.In: Barus V. & O. Oliva (Editors), 1995 Mihulovci Petromyzontes a ryby Osteichthyes (1). (Lampreys Petromyzontes and bony fishes Osteichthyes), Fauna Cˇ R a SR, Vol. 28/1, Academia, Praha, Czech Republic.

Holcik, J., 2006: Is naturalization of paddlefish in the Danube river basin possible? Journal of Applied Ichthyology 22 (Suppl. 1), pp. 40 - 43.

Honsig- Erlenburg, W.; Friedl, M., 1999: Zum Vorkommen des Sterlets (*A. ruthenus* L.) in Kärnten. (Occurence of the sterlet (*A. ruthenus* L.) in Carynthia). Österreichs Fischerei, 52, pp. 129 - 133. [In German].

Honsig- Erlenburg, W.; Petutschnig, W., 2002: Fische, Neunaugen, Flusskrebse, Grossmuscheln. (Fish, lampreys, crayfish and freshwater mussels). Natur Kärnten, Sonderreihe des Naturwissenschaftlichen Vereins für Kärnten Vol. 1, Klagenfurt, Austria, pp. 50 – 51. [In German].

Hugo, A., 1886: Jagdzeitung 29. Retrieved form the archive of Dr. Gertrud Haidvogl, Institute of Hydrobiology and Aquatic Ecosystem Management, University of Natural Resources and Life Sciences, Vienna, Austria. [In German]

Ireland, S.C.; Beamesderfer, R.C.P.; Paragamian, V.L.; Wakkintn, V.D.; Siple, J.T., 2002: Success of hatchery - reared juvenile white sturgeon (*Acipenser transmontanus*) following release in the Kootenai River, Idaho, USA. Journal of Applied Ichthyology 18, pp. 642 - 650.

IUCN/SCC Re - Introduction Specialist Group, 1995: IUCN/SCC Guidelines for Re - Introductions. International Union for the Conservation of Nature, Glad, Pages 1 - 7.

Jäckel, A. J., 1864: Die Fische Bayerns. (The fish of Bavaria). Ein Beitrag zur Kenntnis deutscher Süßwasserfische, Regensburg. Germany, 104 pp. [In German].

Jäckel, A. J., 1866: Ichtyologisches aus meinem Tagebuch 1865. (Ichthyolgical notes of my diary of 1865). Correspondenzblatt des zoologisch-mineralogischen Vereines in Regensburg, 20, (5 – 6), 26 pp. [In German].

Jagsch, A., 1996: Löffelstör in der Aschach? (Paddlefish in the Aschach creek?). Mitteilungen des OÖ LFV. Dezember 1996. OÖ Landesfischereiverband, Linz, Austria. [In German]

Jeitteles, L. H., 1864: Die Fische der March bei Olmütz. (Fishes of the March River near Olmütz). aus: Jahresbericht über das kaiserlich.- königliche Gymnasium in Olmütz während des Schuljahres 1864. 5 pp. [In German].

Jungwirth, M.; Schmutz, S.; Waidbacher, H., 1989: Fischökologische Fallstudie Inn- Fischerbiologische Untersuchung im Hinblick auf Bewirtschaftungsfragen. (Case study on fish ecology in the Inn River). Fischerei- Revierausschuß Innsbruck Stadt und Land, Austria 93 pp. [In German].

Jungwirth, M.; Haidvogl, G.; Moog. O.; Muhar, S.; Schmutz, S., 2003: Angewandte Fischökologie an Fließgewässern. (Applied fish ecology in rivers). Facultas Verlags und Buchhandels AG, Vienna, Austria. pp. 206 - 209. (ISBN 3-8252-2113-X) [In German].

Kärnter Landesregierung, 2001: Verordnung der Landesregierung vom 6. März 2001, Zahl: -11-FIAG 23/2-2001, über die Schonzeiten und Mindestfangmaße (Brittelmaße) für Wassertiere (Kärntner Fischereischonzeitenverordnung - K-FSV). (Size limits and closed seasons for water animals in Carynthia). Klagenfurt. Austria. [In German].

Kerschner, T., 1956: Der Linzer Markt für Süßwasserfische insbesondere in seiner letzten Blüte vor dem ersten Weltkrieg. (The market in Linz for freshwater fish during its peak time before World War one). Naturkundliches Jahrbuch der Stadt Linz 1956. Linz, Austria, pp. 119 – 155. [In German].

Khin A., 1957: A magyar vizák története. (The history of the sturgeons in Hungary). Mezo´´gazdasagi Muzeum Fuzetei 2, pp. 1 - 36. [In Hungarian]

Khodorevskaya, G.; Ruban, G.I.;   Pavlov, D.S., 2009: Behaviour, migrations, distribution and stocks of sturgeons in the Volga- Caspian basin.   World Sturgeon Conservation Society: Special Publication n° 3,. VI, 233 pp. (ISBN 978-3-8391-5449-6)

Khorosko, P.N.; Vlasenko, A.D., 1970: Artificial spawning grounds for sturgeons. Journal of Applied Ichthyology 10, pp. 286 - 292.

Kinzelbach, R., 1994: Ein weiterer alter Nachweis des Sterlet in der Baden-württembergischen Donau. (Another historic catch of sterlet in the Danube in Baden – Würtemmberg). In: Kinzelbach, R., (Editor), 1994: Biologie der Donau. Reihe Limnologie aktuell Vol. 2. Stuttgart, Jena, New York G. Fischer, 370 pp. (ISBN 3–437–30671–5) [In German].

Kirschbaum, F., 2010: Störe – Eine Einführung in Biologie, Systematik, Krankheiten, Wiedereinbürgerung, Wirtschaftliche Bedeutung. (Sturgeons – An Introduction to biology, family tree, diseases, reintroduction and economy). Aqualog animalbook GmbH, Rodgau, Germany, 168 pp. (ISBN 978-3-939759-23-2). [In German].

Kirschbaum, F.; Williot, P.; Fredrich, F., Tiedemann R.; Gessner, J., 2011: Chapter 21 - Restoration of the European Sturgeon *Acipenser sturio* in Germany. pp. 309 – 333. In: Williot, P.; Rochard, E.; Desse - Berset, N.; Kirschbaum, F.; Gessner, J., (Eds.), 2011: Biology and Conservation of the European Sturgeon Acipenser sturio L. 1758 - The Reunion of the European and Atlantic sturgeons. Heidelberg, Dordrecht, London, New York: Springer Verlag. VIII, 668 pp. (ISBN: 987-3-642-20610-8)

Kiwek, F., 1995: Sterlet Programm. Österreichs Fischerei 48, pp. 2 - 3. [In German].

Kottelat, M.; Freyhof, J., 2007: Chapter Family Acipenseridae – sturgeons. pp. 46-60. In: Handbook of European Freshwater Fishes, Kottelat Publications, Cornol, Swisse. XIII, 646 pp. (ISBN 978-2-8399-0298-4)

Kotenev, B. N., 2009: Chapter 21 - Hydrological and Production Characteristics of the Main Basins for Reproduction and Fattening of Sturgeons. pp. 345 - 357. In: Carmona, R.; Domezain, A.; Garcia - Gallego, M.; Hernando, J. A.; Rodriguez F.; Ruiz – Rejon, M., (Eds), 2009: Biology, Conservation and Sustainable Development of Sturgeons. Heidelberg, Dordrecht, London, New York: Springer Verlag. VIII, 668 pp. (ISBN: 987-3-642-20610-8)

Kraft, C., 1874: Die neuesten Erhebungen über die Zustände der Fischerei in den im Reichsrate vertretenen Königreichen und Ländern. (Recent sensus on the state of fishery in the kingdoms and countries represented in the council). Mitt. a. d. Gebiete d. Statistik Jg. XX/IV, Vienna, Austria. [In German].

Krieger, J.; Hett, A.K.; Fuerst, P.A.; Artyukhin, E.; Ludwig, A., 2008: The molecular phylogeny of the order Acipenseriformes revisited. Journal of Applied Ichthyology, 24, pp. 36 - 45.

Krisch, A., 1900: Der Wiener Fischmarkt. Druck und Commissionsverlag von Carl Gerold`s Sohn, Vienna, Austria, 26 pp. Retrieved form the archive of Dr. Gertrud Haidvogl, Institute of Hydrobiology and Aquatic Ecosystem Management, University of Natural Resources and Life Sciences, Vienna, Austria. [In German].

Lagutov, V. (Editor), 2008: Rescue of sturgeon Species in the Ural River Basin. Proceedings of the NATO Advanced Research Workshop on Rescue on Sturgeon Species by means of Transboundary Integrated Water Management of the Ural River Basin, Orenburg, 2007. Heidelberg, Dordrecht, London, New York: Springer Verlag, 333 pp. (ISBN 978-1-4020-8923-7)

Lagutov, V.; Lagutov, V., 2008: The Ural River sturgeons: Population dynamics, catch, reasons for decline and restoration strategies. pp. 193 - 276. In: Lagutov, V. (Editor), 2008: Rescue of sturgeon Species in the Ural River Basin. Proceedings of the NATO Advanced Research Workshop on Rescue on Sturgeon Species by means of Transboundary Integrated Water Management of the Ural River Basin, Orenburg, 2007. Heidelberg, Dordrecht, London, New York: Springer Verlag, 333 pp. (ISBN 978-1-4020-8923-7)

LeBreton, G.T.O.; Beamish, F.W.H.; McKinley, R.S., (Eds.), 2004: Sturgeons and Paddlefish of North America. Dordrecht/Boston/London: Kluwer Academic Publishers, XII, 323 pp. (ISBN 1-4020-2832-6). (Fish and Fisheries Series, vol. 27).

Liepolt, R. (Editor), 1967: Limnologie der Donau. (Limnology of the Danube River). E. Schweizerbart`scheVerlagsbuchhandlung, Stuttgart, Germany, 648 pp. (ISBN 978-3510990528) [In German]

Lori, 1871: Die Fische in der Umgegend von Passau. (The fish in the area of Passau). Beiträge zur Fauna Niederbayerns, Passau, Germany, pp. 99 – 104. [In German].

Ludwig, A.; Lippold, S.; Debus, L.; Reinartz, R., 2009: First evidence of hybridization between endangered sterlets (*Acipenser ruthenus*) and exotic Siberian sturgeons (*Acipenser baerii*) in the Danube River. Biol. Invasions 11, pp. 753 - 760.

Mahen, J., 1927: Castecna revise ryb dunajske obiasti. (Revise of the Danube fish fauna). Sbornik Klubu Pirodovedeckeho v Brne, Vol. 9. 8 pp. [In Czech].

Maier, H. R., 1908: Sterlet im Inn. (Sterlet in the Inn River). Allgemeine Fischerei- Zeitung, 1908, Jrg. 33, Nr. 23, pp.96 - 97. [In German].

Maltsev, S. A., 2009: Chapter 16 - Conservation of the Sturgeon Fish in Lower Volga. pp. 265 - 273. In: Carmona, R.; Domezain, A.; Garcia - Gallego, M.; Hernando, J. A.; Rodriguez F.; Ruiz – Rejon, M., (Eds), 2009: Biology, Conservation and Sustainable Development of Sturgeons. Heidelberg, Dordrecht, London, New York: Springer Verlag. VIII, 668 pp. (ISBN: 987-3-642-20610-8)

Marchetti, M.P.; Nevitt, G.A., 2003: Effects of hatchery rearing on the brain structures of rainbow trout, *Oncorhynchus mykiss*. – Environmental Biology of Fishes 66, pp. 9 – 14.

Margreiter, H., 1927: Ein Sterlet im Inn gefangen. (A sterlet caught in the River Inn). Der Tiroler Fischer 2, pp. 94. [In German].

Mikschi, E.; Wolfram, G., 2007: Rote Liste der Fische (Pisces) Österreichs. (Red list of Austrian fishes). pp. 107 – 136. In: Bundesministerium für Land- und Forstwirtschaft 2007: Rote Listen gefährdeter Tiere Österreichs, Teil 2: Kriechtiere, Lurche, Fische, Nachtfalter, Weichtiere. Böhlau Verlag, Wien-Köln-Weimar, Wien, 515 pp. (ISBN 978-3-205-77478-5) [In German].

Miller, M. J., 2004: Chapter 4 - The Ecology and functional Morphology of Feeding of North American Sturgeon and Paddlefish. pp. 87 - 102. In: LeBreton, G.T.O.; Beamish, F.W.H.; McKinley, R.S., (Eds.), 2004: Sturgeons and Paddlefish of North America. Dordrecht/ Boston/London: Kluwer Academic Publishers, XII, 323 pp. (ISBN 1-4020-2832-6). (Fish and Fisheries Series, vol. 27).

Mohr, E., 1952: Der Stör. (The sturgeon). Akademische Verlagsgesellschaft Geest und Portig K.G., Leipzig, Germany (Die Neue Brehm-Bücherei), 65 pp. Licence No. 276-105/20/32. [In German].

Mojsisovics, A. v. M., 1897: Das Thierleben der österreichisch-ungarischen Tiefebenen. (Animal life in the Austrian-Hungarian planes). Alfred Hödler, Vienna, Austria. Retrieved form the archive of Dr. Gertrud Haidvogl, Institute of Hydrobiology and Aquatic Ecosystem Management, University of Natural Resources and Life Sciences, Vienna, Austria. [In German].

NÖ. Landesregierung, 2002: Niederösterreichische Fischereiverordnung 2002. (Fishery regulations in lower Austria). St. Pölten, Austria, Page 6. [In German].

Nordheim, H. v.; Gessner, J.; Kirschbaum, F., Anders, E. & G. M. Arndt, 2001: Das Wiedereinbürgerungsprogramm für *A. sturio* - Hintergründe und Konzeption. pp. 30 – 49. In: Anonymous, 2001: Der Stör Acipenser sturio L. Fisch des Jahres 2001. (The sturgeon: fish of the year 2001). Verband deutscher Sportfischer e. V. (editor), Offenbach, Germany, 86 pp. [In German].

OÖLFV (Oberösterreichischer Landesfischereiverband), 1996: Mitteilungen des O.Ö. (Ober-öster-eischichen)LFV (Landesfischerei-Verband). Juni 1996, Linz, Austria, pp. 228. [In German].

OÖLFV (Oberösterreichischer Landesfischereiverband), 1997: Leitfaden zur Fischkunde und Angelfischerei. (Guideline to fish taxonomy and sportfishing). Linz, Austria, 384 pp. [In German].

OÖLFV (Oberösterreichischer Landesfischereiverband), 2004: Fischerei und Gesetz. (Upper Austrian fishing regulations). Linz, Austria. Page 72. [In German].

Österreichs Fischereiwirtschaft, 1936: Fang eines großen Störes in Wien. (Catch of a large sturgeon in Vienna.). IV Jahrgang, Page 69. [In German].

Österreichischer Fischereiverband, 2005: Waxdick- Besatz für die Donau. (Stocking of Russian sturgeon into the Danube). Österreichs Fischerei, 58, Pages 193 - 194. [In German].

Peake, S. J., 2004: Chapter 7 - Swimming and Respiration. pp. 147 – 166 In: LeBreton, G.T.O.; Beamish, F.W.H.; McKinley, R.S., (Eds.), 2004: Sturgeons and Paddlefish of North America. Dordrecht/Boston/London: Kluwer Academic Publishers, XII, 323 pp. (ISBN 1-4020-2832-6). (Fish and Fisheries Series, vol. 27).

Pegasov, V. A., 2009: Chapter 20 - The Ecological Problems of Introduction and Reintroduction of Sturgeons. pp. 339 – 343. In: Carmona, R.; Domezain, A.; Garcia - Gallego, M.; Hernando, J. A.; Rodriguez F.; Ruiz – Rejon, M., (Eds), 2009: Biology, Conservation and Sustainable Development of Sturgeons. Heidelberg, Dordrecht, London, New York: Springer Verlag. VIII, 668 pp. (ISBN: 987-3-642-20610-8)

Peyrer, C., 1874: Fischereibetrieb und Fischereirecht in Österreich. (Fisheries and fishing regulations in Austria). Ackerbauministerium, Vienna.,Austria. Retrieved form the archive of Dr. Gertrud Haidvogl, Institute of Hydrobiology and Aquatic Ecosystem Management, University of Natural Resources and Life Sciences, Vienna, Austria. [In German].

Pinter, K., 2008: Rearing and Stocking of Brown Trout, *Salmo trutta L.:* Literature Review and Survey of Austrian Fish Farmers within the Frame of the Project - Initiative Troutcheck. Department of Water, Atmosphere and Environment - Institute of Hydrobiology and Aquatic Ecosystem Management, University of Natural Resources and Applied Life Sciences, Vienna, Austria, D- 13631, 120 pp.

Puzzi, C. M.; Bellani, A.; Trasforini, S.; Ippoliti, A., 2009: Chapter 17 - Experience of Conservation of *Acipenser naccarii* in the Ticino River Park (Northern Italy). pp. 275 – 298. In: Carmona, R.; Domezain, A.; Garcia - Gallego, M.; Hernando, J. A.; Rodriguez F.; Ruiz – Rejon, M., (Eds), 2009: Biology, Conservation and Sustainable Development of Sturgeons. Heidelberg, Dordrecht, London, New York: Springer Verlag. VIII, 668 pp. (ISBN: 987-3-642-20610-8)

Radovanovic, I., 1997: The Lepenski Vir culture: A contribution to interpretation of its ideological aspects. pp. 87-93. In: M. Lazic (Editor) Antidoron Dragoslavo Srejovic completis LXV annis ad amicis, collegis, discipulis oblatum. Belgrade: Center for Archaeological Research, pp. 87 – 93.

Reichle, G., 1997: Der Stör. (The sturgeon). Verlag Lassleben, Kallmünz, Germany, 80 pp. (ISBN 3-7847-9115-8) [In German].

Reinartz, R., 2008: Artenhilfsprogramm Sterlet- Abschlussbericht 2004 - 2007. (Restoration programme sterlet – final report 2004 – 2007) Landesfischereiverband Bayern, Munich, Germany, 53 pp. [In German].

Remes, 1902: Ryby moravske. (Fish of the Morava River). Cas.Vlast.sp.mus, Olomouc, Czech Republic, 62 pp. [In Czech].

Rosenthal, H.; Bronzi, P.; McKenzie, D.J.; Arlati, G.; Rossi, R., (Editors), 1997: Proceedings of the 3rd International Symposium on Sturgeons, Piacenza, Italy. Journal of Applied Ichthyology 15 (4-5), 352 pp.

Rosenthal, H.; Bronzi, P.; Spezia, M., Poggioli, C; (Editors), 2008: Passages for fish - overcoming barriers for large migratory fish. World Sturgeon Conservation Society, Special Publication No. 2, IV, Books on Demand GmbH, Norderstedt, Germany, 165 pp. (ISBN – 978-3-8370-6142-0).

Salzburger Landesregierung, 2003: Salzburger Fischereiverordnung, (Salzburg Fisheries decree). LGBl. Nr. 1/2003 §1. Austria. [In German].

Schmall B.; Ratschan, C., 2011: Die historische und aktuelle Fischfauna der Salzach - ein Vergleich mit dem Inn. (Historic and present fish fauna of the river Salzach – a comparison with the river Inn). Beitrag Naturkunde Oberösterreichs 21, pp. 55 - 191. [In German].

Siebold, C.T.E., 1863: Die Süßwasserfische von Mitteleuropa. (The freshwater fishes of Central Europe). Verlag von Willhelm Engelmann, Leipzig, Germany, 452 pp. [In German].

Simonovic, P.; Maric, S.; Nikolic, V., 2006: Occurrence of paddlefish *Polyodon spathula* (Walbaum, 1792) in the Serbian part of the lower River Danube. Aquatic Invasions 1, (3), pp. 183 - 185.

Spindler, T., 1994: Status der Fischfauna der March. (State of the fish fauna in the river March). Wissenschaftliche Mitteilungen des Niederösterreichischen Landesmuseums 8, pp. 177 – 189. [In German].

Spindler, T., 1997 : Fischfauna in Österreich. (Fishfauna of Austria). Monographien Band 87, Bundesministerium für Umwelt, Jugend und Familie, Vienna, Austria, 157 pp. (ISBN 3-85457-217-4) [In German]

Steiermärkische Landesregierung, 2000: Verordnung der Steiermärkischen Landesregierung vom 11. Dezember 2000 über die Schonzeiten und Mindestfanglängen von Wassertieren. (Regulation of the state authorities of the Steiermark as of 11. December 2000 on closed seasons and minimum catch length of aquatic organisms). LGBl. Nr. 81/2000, Graz, Austria, pp. 1. [In German].

Streibl, D., (without Date): Über den Sterlet. (About the sterlet). Cited in: Reinartz, R., 2008: Artenhilfsprogramm Sterlet- Abschlussbericht 2004 - 2007. (Restoration programme sterlet – final report 2004 – 2007) Landesfischereiverband Bayern, Munich, Germany, 53 pp. [In German].

Sulak, K.J.; Clugston, J.P., 1999: Early life history and population structure of Gulf sturgeon, *Acipenser desotoi*, in the Suwanee River, Florida. Journal of Applied Ichthyology 15, pp. 116 - 128.

Van Eeneennaam, J.P.; Chapman, F.A.; Jarvis, P.L., 2004: Chapter 13 - Aquaculture. pp. 300. In: LeBreton, G.T.O.; Beamish, F.W.H.; McKinley, R.S., (Eds.), 2004: Sturgeons and Paddlefish of North America. Dordrecht/Boston/London: Kluwer Academic Publishers, XII, 323 pp. (ISBN 1-4020-2832-6). (Fish and Fisheries Series, vol. 27).

Wacha, G., 1956: Fische und Fischhandel im alten Linz. (Fish and fish trade in the old Linz). Naturkundliches Jahrbuch der Stadt Linz. Austria, 57 pp. [In German].

Waidbacher, H.; Zauner, G.; Kovacek, H.; Moog, O., 1991: Fischökologische Studie Oberes Donautal in Hinblick auf Strukturierungsmaßnahmen im Stauraum Aschach. (Fish ecological study of the upper Danube Valley with regard on restructuring measures in the impoundment Aschach). Department of Water, Atmosphere and Environment - Institute of Hydrobiology and Aquatic Ecosystem Management, University of Natural Resources and Applied Life Sciences, Vienna, Austria, 175 pp. [In German].

Waidbacher, H.; Straif, M.; Drexler, S. 2006: Erhebung und Einschätzung des Erhaltungszustandes der in Anhang II und V der FFH - Richtlinie genannten und in Wien vorkommenden und geschützten Fischarten - Berichtsjahr 2006. (Evaluation of the status of protected fish species in Vienna listed in the appendices II and V of the FFH – Directive). Department of Water, Atmosphere and Environment - Institute of Hydrobiology and Aquatic Ecosystem Management, University of Natural Resources and Applied Life Sciences, Vienna. Austria, 82 pp. [In German].

Weeger, 1884: Fische von Donau und Mähren. (Fish of Danube and Mähren). Mitteilungen des Österreichischen Fischereivereins, 1887. Retrieved form the archive of Dr. Gertrud Haidvogl, Institute of Hydrobiology and Aquatic Ecosystem Management, University of Natural Resources and Life Sciences, Vienna, Austria. [In German].

Wiener Landesregierung, 2008: 44. Verordnung der Wiener Landesregierung betreffend Schonzeiten und Mindestmaße der Fische sowie Krebse und Muscheltiere. (Regulations of the Viennese government on minimum catch length and closed seasons on fish crayfish and mussels). Vienna, Austria, pp. 2. [In German].

Wiesner, C.; Wolter, C.; Rabitsch, W.; Nehring, S., 2010: Gebietsfremde Fische in Deutschland und Österreich und mögliche Auswirkungen des Klimawandels. (Alien fish species in Germany and Austria and possible effects of the climate change). Bundesamt für Naturschutz, Bonn - Bad Gobesberg, 192 pp. (ISBN 978-3-89624-014-9) [In German].

Williot, P.; Rochard, E.; Kirschbaum, F., 2009: Chapter 23 - Acceptability and Prerequisites for the Successful Introduction of sturgeon species. pp. 369 – 384. In: Carmona, R.; Domezain, A.; Garcia - Gallego, M.; Hernando, J. A.; Rodriguez F.; Ruiz – Rejon, M., (Eds), 2009: Biology, Conservation and Sustainable Development of Sturgeons. Heidelberg, Dordrecht, London, New York: Springer Verlag. VIII, 668 pp. (ISBN: 987-3-642-20610-8)

Williot, P.; Rochard, E.; Desse - Berset, N.; Kirschbaum, F.; Gessner, J., (Eds.), 2011: Biology and Conservation of the European Sturgeon Acipenser sturio L. 1758 - The Reunion of the European and Atlantic sturgeons. Heidelberg, Dordrecht, London, New York: Springer Verlag. VIII, 668 pp. (ISBN: 987-3-642-20610-8).

Windeck, E., 1445 - 1450: Kaiser Sigismunds Buch. Retrieved from the archive of Dr. Gertrud Haidvogl, Institute of Hydrobiology and Aquatic Ecosystem Management, University of Natural Resources and Life Sciences, Vienna, Austria. [In German].

Woschitz, G., 2006: Rote Liste gefährdeter Fische in der Steiermark. (Red list of endangered fish species in Styria.). Unpublished study for the Government of Styria, Austria. [In German].

Zauner, G., 1997: Acipenseriden in Österreich. (Acipenserids in Austria). Österreichs Fischerei, 50, pp. 183 - 187. [In German].

Zauner, G.; Pinka, P.; Jungwirth, M., 2000: Wasserwirtschaftliches Grundsatzkonzept Grenzmur - Phase 1, TB 2.1 Fischökologie. (Water management plan for the Mura River along the Austro-Slovenian Border- Part 2.1 fishecology). Fachabteilung 3a, Eigenverlag EZB, Engelhartszell, Austria. [In German].

Zauner, G.; Pinka, P.; Moog, O., 2001: Pilotstudie Oberes Donautal: Gewässerökologische Evaluierung neugeschaffener Schotterstrukturen im Stauwurzelbereich des Kraftwerks Aschach. (Pilot study Upper Danube Valley – evaluation of reconstructed gravel banks in the headwater of the impoundment Aschach). Department of Water, Atmosphere and Environment - Institute of Hydrobiology and Aquatic Ecosystem Management, University of Natural Resources and Applied Life Sciences, Vienna, Austria, 133 pp. [In German].

Zborˇil, J.; Absolon, K., 1916: Zoologicka´ pozorova´nı´ z okolı´ hodonı´nske´ho (Zoological observations from the Hodonın region). Cˇ asopis moravske´ho musea zemske´ho 15, pp. 3 - 12. [In Czech].

Zetter, J.T.M., 1862: Vortrag über die Einbürgerung der künstlichen Fischzucht in Salzburg. (Lecture about the establishment of hatcheries in Salzburg) Monats-Blatt der k.k. Landwirtsch.-Gesellschaft. Salzburg 12, pp. 19 – 53. [In German].

Retrieved via internet

Angelforum: www.angelforum.at/storfange-in-osterreichischen-flieszgewassern-t10941.html
accessed last 25.05.2013

Donaufischer: www.donaufischer.at/module-Fischerforum-viewtopic-topic-1631.phtml
accessed last 17.08.2012

Fischerforum:
www.fischerforum.or.at/index.php?page=Thread&postID=54559&highlight=sterlet#post54559 accessed last 17.08.2012

www.fischerforum.or.at/index.php?page=User&userID=4368
accessed last 17.08.2012

www.fischerforum.or.at/index.php?page=Thread&threadID=8792
accessed last 17.08.2012

Hechtsprung: www.hechtsprung.tv
accessed last 04.10.2006

Homepage of the Fischereiverein Leibnitz: www.fvl.at/cms/
26.11.2010; accessed last 06.04.2012

Homepage of the village Hofkirchen: www.hofkirchen.at/artikel.shtml?newsid=2521
accessed last 06.04.2012

Homepage of the Landesfischereiverband Oberösterreich:
www.lfvooe.at/Reviere/donau_rohrbach_loeffel.html
accessed last 22.04.2013

Homepage of the Wiener Fischereiausschuss:
www.wiener-fischereiausschuss.at/huchen_sterlet.htm
accessed last 2204.2013

Homepage of the World Sturgeon Conservation Society:
www.wscs.info - Bulgaria issued ban on sturgeon fishing, 18.03.2011
accessed last 25.05.2013

Figure A1: White sturgeon (A. transmontanus) caught in the mouth of the river Salzach around the year 2002. (unknown)

Störfall am Stammhamer Kraftwerk

Fig. A2

Figure A2 and A3: Newspaper articles on the catch of a "sturgeon" (Fig. 2) and three "sterlets" (Fig. 3) in the river Inn near Burghausen in 2004. The species caught are actually one Russian sturgeon (A. gueldenstaedtii) and two Siberian sturgeons (A. baerii) and one specimen of unknown species. (provided by FV Burghausen, Burghausen, Germany)

Ein äußerst seltener Fang

Am Freitag dem 18. Juli 2003 ging unserem Vereinskamerad Johann Wimmer ein wahrlicher kurioser Fisch an die Angel. Der Tauwurm, den er morgens um 9.00 Uhr an der Alzmündung ausgelegt hatte, war eigentlich für Forellen gedacht.

Johann Wimmer staunte allerdings nicht schlecht als nach dem Biß statt einer „Regenbognerin" ein Sterlett von 82 cm Länge an der Angel hing. Gleich nach dem Anhieb sprang der Sterlett aus dem Wasser, wie sonst nur Forellen, danach aber ging er auf Tiefgang und verhielt sich beim Drill ähnlich störrisch wie die Barben, nur lange nicht so ausdauernd. Doch jetzt begann die Sache erst interessant zu werden. Darf der Fisch entnommen werden, hat er ein Schonmaß oder eine Schonzeit?

So viele Fragen und keine Antworten. Johann Wimmer hälterte den seltenen Fang in einem Granter und versuchte telefonisch Klarheit zu bekommen. Nach einigen Versuchen jemand von der Vorstandschaft zu erreichen hatte er bei unserem Vorsitzenden Wolfgang Schneidermeier endlich Glück. Er wußte zwar auch nicht ganz sicher ob der Fisch entnommen werden darf, aber er konnte soweit Auskunft geben, dass es schon einmal einen ähnliche Fall gab bei dem der Fang wieder zurück ins Gewässer kam.

Daraufhin wurde er nach einem schönen Foto wieder in die Fluten des Inns schonend entlassen, und wer weiß vielleicht beißt er ja auch mal bei einem von uns an.

Noch schnell ein Foto und dann wieder zurück in den kühlen Inn und somit in die Freiheit. (Der Sterlett ist im übrigen in unserem Vereinsgewässer ganzjährig geschont)

Fig. A3

Figure A4: The Russian sturgeon described in the Newspaper article shown in Figure 3, documenting the wrong species identification in the media presentation. (provided by Christian Zagler, Burghausen, Germany)

Figure A5: Two sterlets (A. ruthenus) (about 50 and 35 cm TL) caught downstream of the Jochenstein power plant in 2002. Note the albino specimen. (provided by TB Zauner, Engelhartszell, Austria)

Figure A6: Albino specimen (about 35 cm TL) of A. ruthenus caught downstream of the Jochenstein power plant in 2002. (provided by TB Zauner, Engelhartszell, Austria)

Figure A7: Sterlet (A. ruthenus) (about 60 cm TL) caught downstream of the Jochenstein power plant in 2002. (provided by TB Zauner, Engelhartszell, Austria)

Figure A8: A large sterlet specimen (A. ruthenus), measuring 76 cm total length, caught downstream of the Jochenstein power plant in 2006. (provided by TB Zauner, Engelhartszell, Austria)

Figure A9: Siberian sturgeon (A. baerii) caught downstream of the Jochenstein power plant in 2006/2007. (provided by TB Zauner, Engelhartszell, Austria)

Figure A10: Sterlet (A. ruthenus) (about 45 cm TL) caught downstream of the Jochenstein power plant in 2007. (provided by TB Zauner, Engelhartszell, Austria)

Figure A11: Sterlet (A. ruthenus) (50 cm TL) caught downstream of the Jochenstein power plant in 2008. (provided by TB Zauner, Engelhartszell, Austria)

Figure A12: Sterlet (A. ruthenus), with around 60 cm total length, caught downstream of the Jochenstein power plant in 2008. (provided by TB Zauner, Engelhartszell, Austria)

Figure A13: Sterlet (A. ruthenus) caught downstream of the Jochenstein power plant in 2011. This specimen was stocked in the German river Schwarze Laber and travelled 150 km downstream. Its total length was 35 cm. (provided by TB Zauner, Engelhartszell, Austria)

Figure A14: Sterlet (A. ruthenus) caught downstream of the Jochenstein power plant in 2011. It was in very good condition and measured 54 cm TL. (provided by TB Zauner, Engelhartszell, Austria)

Figure A15: Sterlet (A. ruthenus) (49 cm TL) caught downstream of the Jochenstein power plant in 2011. Note the untypical number of dorsal plates. (provided by TB Zauner, Engelhartszell, Austria)

Figure A16: Sterlet (A. ruthenus) (46 cm TL) caught downstream of the Jochenstein power plant in 2011. (provided by TB Zauner, Engelhartszell, Austria)

Figure A17: Sterlet (A. ruthenus) (49 cm TL) caught downstream of the Jochenstein power plant in 2011. (retrieved from TB Zauner, Engelhartszell, Austria)

Figure A18: Sterlet (A. ruthenus), with around 45 cm total length, caught downstream of the Jochenstein power plant in 2011. (retrieved from TB Zauner, Engelhartszell, Austria)

Figure A19: Juvenile sterlet (A. ruthenus), with a total length of around 30 cm, caught downstream of the Jochenstein power plant in 2011. (retrieved from TB Zauner, Engelhartszell, Austria)

Figure A20: Large specimen of sterlet (A. ruthenus), measuring 76 cm total length, caught near Niederranna, Upper Austria in 2005. (retrieved from TB Zauner, Engelhartszell, Austria)

Figure A21: Paddlefish (Polyodon spathula) (115 cm TL, 8.7 kg) caught near Schlögen in 2003. http://www.lfvooe.at/Reviere/donau_rohrbach_loeffel 26.05.2003

Figure A22: Small sterlet (A. ruthenus) caught near Linz in 1984, long before stocking actions took place. (provided by Franz Lahmer, Linz, Austria)

Figure A23: Sturgeon claimed to be a "Hybrid" caught near Linz around the year 2002. Possibly Acipenser naccarii x Huso huso according to reference source. (cited from ANONYMOUS, 2005)

Figure A24: The same hybrid variant (about 110 cm TL) as above caught downstream of Linz in the year 2003. (unknown source, Linz, Austria)

Figure A25: Siberian sturgeon (A. baerii), around 75 cm total length, caught in the harbour of Linz in 1999. (provided by Franz Wiesmayer, Linz, Austria)

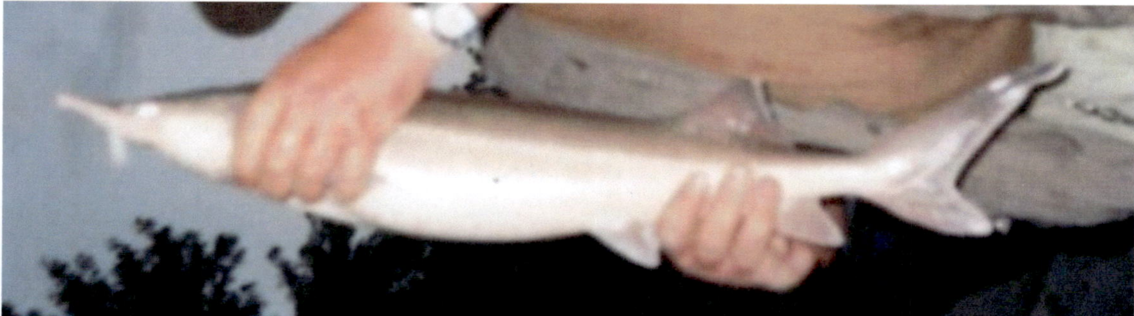

Figure A26: Sterlet (A. ruthenus) (50 cm TL) caught downstream of the power plant Wallsee - Mitterkirchen in 1997. (provided by Hans- Peter Angerer, Austria)

Figure A27: Newspaper clip of an article describing stocking of "Beluga sturgeon" in Linz in 1996. The actual species of the stocked fish shown is Acipenser gueldenstaedtii, the Russian sturgeon. (retrieved from Siegfried Pilgerstorfer, OÖ Landesfischereiverband, Austria)

Records from the Danube in Lower Austria

Stör in der Donau

WAS MACHT ein Stör in der Donau? Diese Frage kann wohl nicht einmal Neptun selbst beantworten. Tatsache jedenfalls ist, dass **Heinz Renner**, seines Zeichens Chef des „Wachauerhofes" in Marbach an der Donau, einen vier Kilo schweren und ein Meter langen Jungfisch dieser Art aus Österreichs großem Strom ziehen konnte. Der Hausen-Beluga-Stör ist normalerweise im Atlantik, in der Nord- oder Ostsee zuhause. Wie er sich in die Donau verirrt hat, bleibt rätselhaft. Einerlei: Den eifrigen Petrijünger freut's! Und auch Enkerl Bettina sowie Schwiegersohn Joachim sind hellauf begeistert.

Figure A28: Newspaper article regarding the catch of a "beluga sturgeon" in Marbach in 2003. The actual species of the fish is A. gueldenstaedtii, the Russian sturgeon. (unknown source)

Figure A29: Very large sterlet specimen (91 cm TL) in good condition, caught downstream of the Melk power plant in 2007. The fish was released alive shortly after the catch. (provided by Christoph Trost, Melk, Austria)

Figure A30: Siberian sturgeon (A. baerii) caught in the Lower Austrian Danube. (precise date and source unknown)

Figure A31: Siberian sturgeon (A. baerii) caught in the Lower Austrian Danube. Note the good condition of the specimen. As the fisherman believed the specimen to be a ship sturgeon it was released alive. (retrieved from Mathias Jungwirth, Vienna, Austria)

Figure A32: Hybrid (possibly A. gueldenstaedtii x Huso huso) caught downstream of the Altenwörth power plant in 2002. (Institute of Hydrobiology and Aquatic ecosystem management, Vienna, Austria)

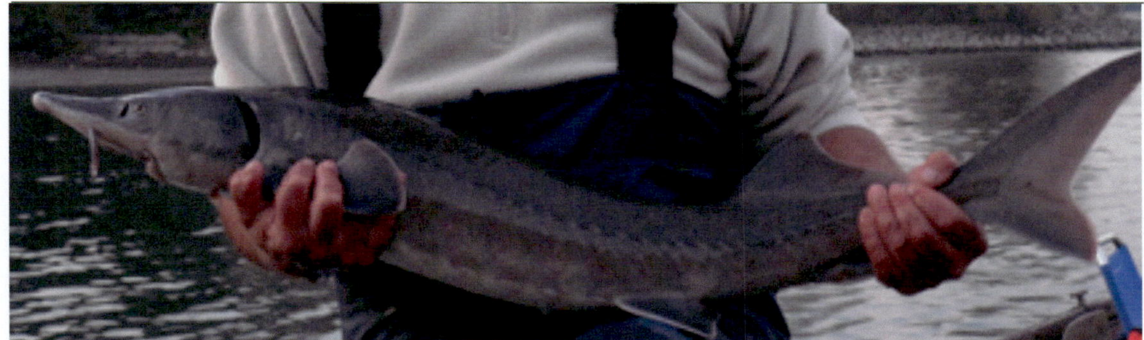

Figure A33: Siberian sturgeon (A. baerii), measuring 96 cm total length, caught downstream of the Altenwörth power plant in 2005. This fish is also in very good condition. (retrieved from Mathias Jungwirth, Vienna, Austria)

Figure A34: Siberian sturgeon (A. baerii) caught downstream of the Altenwörth power plant in 2011. (retrieved from Mathias Jungwirth, Vienna, Austria)

Figure A35: Large sterlet (A. ruthenus) caught downstream of the Altenwörth power plant in 2003. Note the deformed body. Due to its size of around 90 cm total length it is probably not stocked but rather a wild specimen. (provided by Robert Elsbacher, Austria)

Figure A36: Sterlet (A. ruthenus) caught downstream of the power plant Freudenau in Vienna in 1999 or 2000. (provided by TB Eberstaller, Vienna, Austria)

Figure A37: Sterlet (A. ruthenus) caught downstream of the power plant Freudenau in Vienna in 2011. This specimen might originate from the stocking programme in the years 2002 – 2006.   (provided by Robert Schlappal, Vienna, Austria)

Figure A38: Sterlet (A. ruthenus) to be stocked into the Freudenau impoundment between 2002 and 2005. (retrieved from Christian Wiesner, Vienna, Austria)

Figure A39: Stocking of sterlets near Regelsbrunn in 1994. (provided by Martin Hochleithner, Kitzbühel, Austria)

Figure A40: Stocking of sterlets in the Danube - Auen National Park in 2001. (provided by Martin Hochleithner, Kitzbühel, Austria)

Figure A41: Stocking of Russian sturgeons in the Danube - Auen National Park in 2005. (provided by Leopold Feichtinger, Wels, Austria)

Figure A42: Stocking of Russian sturgeons in the Danube - Auen National Park in 2005. (provided by Leopold Feichtinger, Wels, Austria)

Figure A43: Stocking of a few specimens of sterlet as public relations-activity in 2010 in Vienna. (obtained from VIA- Donau Waterway management Austria), Vienna, Austria)

Records from other tributaries

Löffelstör in der Aschach?

Das Institut für Gewässerökologie, Fischereibiologie und Seenkunde, Scharfling, wurde informiert, daß Anfang November in der Aschach (Bereich Eferding) ein toter Löffelstör an einem Rechenbauwerk aufgefunden wurde. Die Heimat des Löffelstörs *(Polyodon spathula)* ist das Gebiet des Mississippi-Missouri in Nordamerika; er kann bis über 1,5 m lang und 70 kg schwer werden, seine Hauptnahrung ist Zooplankton. **Der Löffelstör ist eine nicht heimische Fischart und darf in Oberösterreich nicht ohne Bewilligung ausgesetzt werden (§ 10 LFG).** Wer gegen diese Bestimmungen verstößt begeht eine Verwaltungsübertretung und hat mit Strafen im Höchstausmaß vor S 30.000,– zu rechnen.

Dr. Albert Jagsch

Figure A44: Newspaper article on a dead paddlefish (Polyodon spathula) found in the river Aschach. (retrieved from Siegfried Pilgerstorfer, OÖ Landesfischereiverband)

Records from Drava River

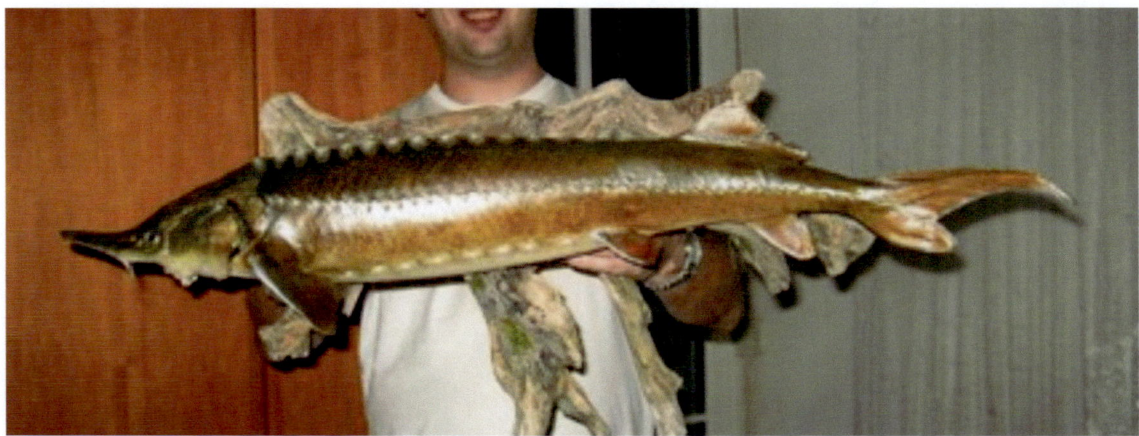

Figure A45: Preparate of a large sterlet (A. ruthenus) (94 cm TL, 4,8 kg) caught near Wunderstätten in 2003. (unknown source)

Records from Mura River

Figure A46: Siberian sturgeon (A. baerii) found dead above the power plant Murau in 2005. (provided by Christian Wiesner, Vienna, Austria)

Figure A47: Siberian sturgeon (A. baerii) caught near Wildon in 2010. (provided by Josef Riedl, Austria)

Figure A48: Siberian sturgeon (A. baerii) caught near Wildon in 2010. (provided by Stefan Pretenhofer, Austria)

Records from natural lakes

Figure A49: Newspaper article from around the year 2000 regarding the catch of a "sterlet" in Lake Hallstatt. The actual species of this specimen is however the Siberian sturgeon (A. baerii) probably released by a private person (aquarist?). (unknown source)

Figure A50: Magazine article regarding the catch of six "sterlets" in the Traunsee. The species on the picture is a Siberian sturgeon (A. baerii), and there is also a report of stocking of ten specimens of A. baerii into the Traunsee (MAIER pers. comm.). (cited from Fisch & Fang 12/2000)

ADDITION:
Records from the Danube in Upper Austria

Figure A51: Russian sturgeon (A. gueldenstaedtii) caught in fall 2012 below the power plant Jochenstein. (provided by TB Zauner, Engelhartszell, Austria)

Annex II - The Aschach Impoundment

As it is the home of the last self - reproducing population of the sterlet (Acipenser ruthenus) the Aschach impoundment shall be briefly characterized: The impoundment has a length of around 42 km and is the longest impoundment within the Austrian Danube. It is partially situated along the Upper Austrian - Bavarian border with its upper boundary, the Jochenstein power plant at km 2203 and the Aschach power plant on its downstream end at km 2162 (Fig. B1). The Jochenstein power plant was built in 1956 and the Aschach power plant in 1964. Gravel banks became submersed and lost most of their ecological functions, while shorelines were stabilized with rip - raps (ZAUNER et al., 2001). The headwater section with similar hydrological conditions to a natural river is around 7 km long (Fig. B2). The stagnant section starts near the Niederranna Bridge (Fig. B2 - on the right) and is around 35 km long (WAIDBACHER et al., 1991). The upper part is characterized through rocky substrates and high flow velocities. Although sediment transport is blocked through various power plants the sediment composition in this area correlates with the natural condition. Further downstream gravel and sandy substrates are dominant. At the end of the 20th century various revitalizations were undertaken including the construction of gravel banks (ZAUNER et al., 2001).

Fig. B1: The impoundment Aschach and the famous Schlögener Schlinge. (Google/Geoimage Austria 2012)

Fig. B2: Upper section of the impoundment with the Jochenstein power plant on the left and the Niederranna Bridge on the right. (Google/Geoimage Austria 2012)